SBus

*Information, Applications,
and Experience*

James D. Lyle

SBus
Information, Applications, and Experience

With 98 Illustrations

Springer-Verlag
New York Berlin Heidelberg London Paris
Tokyo Hong Kong Barcelona Budapest

James D. Lyle
Troubador Technologies
P.O. Box 2606
Santa Clara, CA 95055-2606
USA

Library of Congress Cataloging-in-Publication Data
Lyle, Jim.
    SBus : information, applications, and experience / Jim Lyle.
        p. cm.
    Includes bibliographical references and index.
    1. SBus (Computer bus) I. Title.
TK7895.B87L95 1992
004.6'4--dc20   92-18701

Printed on acid-free paper.

Production managed by Karen Phillips; manufacturing supervised by Jacqui Ashri.
Photocomposed pages prepared from the author's FrameMaker file.
Printed and bound by R.R. Donnelley & Sons, Harrisonburg, VA.
Printed in the United States of America.

9 8 7 6 5 4 3 2 1

ISBN 0-387-97862-3 Springer-Verlag New York Berlin Heidelberg
ISBN 3-540-97862-3 Springer-Verlag Berlin Heidelberg New York

Acknowledgments

I have collected the information in this book from a wide variety of sources. These are acknowledged throughout the book, at the end of each chapter and in the bibliography. Special acknowledgements, however, must be given to many people who contributed information, ideas, suggestions and (where necessary) emotional support. Their influence is so pervasive that it would be impossible to acknowledge every instance, just as it would have been impossible to write this book without their help.

First, I'd like to thank my fellow alumni of Sun Microsystem's SBus Technical Support Group (now disbanded), including Jim Lockwood, Mike Saari, Deborah Bennett, and others. I'd also like to thank the members of the SBus Specification Committee and the (IEEE) P1496 Working Group. Too numerous to mention, these professionals taught me much about the intricacies and compromises of bus design in general, and SBus in particular.

I'd also like to thank Michelle Gutierrez, who spent many hours reviewing and editing the material, and Ko Yamamoto, Brad Giffel, and all the other engineers who helped review the book for technical accuracy.

Most importantly, though, my warmest regards and sincerest thanks go to Barbara Vance. She took most of the photographs this book contains, and did significant research, review, and editing. I am especially grateful for the dinners she brought me all those nights I worked late.

May, 1992

James D. Lyle

Contents

Introduction

Workstation and computer users have an ever increasing need for solutions that offer high performance, low cost, small footprints (space requirements), and ease of use. Also, the availability of a wide range of software and hardware options (from a variety of independent vendors) is important because it simplifies the task of expanding existing applications and stretching into new ones. The SBus has been designed and optimized within this framework, and it represents a next-generation approach to a system's I/O interconnect needs.

This book is a collection of information intended to ease the task of developing and integrating new SBus-based products. The focus is primarily on hardware, due to the author's particular expertise, but firmware and software concepts are also included where appropriate.

This book is based on revision B.0 of the SBus Specification. This revision has been a driving force in the SBus market longer than any other, and is likely to remain a strong influence for some time to come. As of this writing there is currently an effort (designated P1496) within the IEEE to produce a new version of the SBus specification that conforms to that group's policies and requirements. This might result in some changes to the specification, but in most cases these will be minor. Most of the information this book contains will remain timely and applicable. To help ensure this, the author has included key information about proposed or planned changes.

It is hoped that this book will be a useful reference, but it is not intended to serve as a replacement for the SBus specification. Please consider the specification the ultimate authority on SBus design, and make every effort to adhere to it.

Why the SBus?

1

Computers and workstations are appearing in a rapidly increasing number of places and types of applications. As technology improves, people who only a few years ago would never have dreamed of using such a machine are now finding them a part of their everyday lives. These people are increasingly non-technical, and have needs that go far beyond raw MIPS and bandwidth.

As applications become more diverse, computer and workstation vendors are finding it impossible to provide proprietary solutions for every customer need. Therefore, third-party hardware and software vendors have become an important part of the computing marketplace by filling in the gaps. This ultimately helps consumers (whose needs are better fulfilled), the third-parties (who profit from their role), and computer vendors (who can now concentrate their resources on the system, rather than the application).

Open, standard interfaces are essential in this type of marketplace, because they are the framework upon which multi-vendor applications are built. The SBus is an example of one such interface, and is designed and optimized to best meet the needs of developers, vendors, and end users.

1.1 Why Design a New Bus?

Before any new design is undertaken, it is important to fully understand the requirements it will be expected to satisfy, and to examine existing solutions. If no suitable match is found, then a new design is necessary.

1.1.1 Key Requirements for New Machines

This section examines the requirements of an I/O interconnect that is suitable for use in a desktop platform, and that can also form part of larger systems. The users of these systems include the traditional technically sophisticated user and the less technically adept business user.

Low Cost

The total cost of a system to the customer includes the initial cost of the base system and the add-on cards, as well as maintenance and service costs. Ideally, to reduce initial costs, a bus should not require expensive circuitry to accomplish simple functions, and will have costs that scale with the demands of the application. Similarly, complex functions, such as bus arbitration, which need not be replicated on each card, should reside in one place in the system. Maintenance and service costs can be reduced by simplifying installation and increasing mean time between failures (MTBF).

Low Power

Low drive power helps reduce power supply size and cooling constraints (of utmost concern in some applications, such as laptops). Also, throughout the development of the SBus, the maximum 4 – 8-milliamp (mA) driver capacity of most of today's large CMOS ASICs was a very important limitation. If more current than this were needed, buffers would be necessary, and that would add significantly to power, costs, and complexity. Instead, SBus has been developed to allow these parts to be connected directly together, to great advantage.

Wide Variety of Applications Available

One key point to remember is that people won't buy a machine just because of the kind of bus it supports. You don't buy an SBus-compatible machine *because* you think its bus is superior (although that might be a factor). What you care about isn't the technical excellence of the interface, but the number and kind of *applications* that you can plug into it. From the end-user's perspective, one of the most important aspects of an expansion bus is the number of options it brings with it. A technically superior bus without many expansion cards is usually of much less value than a bus that is technically inferior, yet has greater industry acceptance and vendor support.

Likewise, as a developer, you don't choose to port an application to one bus or the other based on this technical issue or that. You are more concerned with how many people will buy your product. The vendors' acceptance of a particular bus depends on their perception of the market size, and their anticipated development schedules, costs, and risks. A bus with an interface that is easy to design will address all of these factors, making the bus a more attractive candidate for porting new applications.

Performance

One measure of performance is the ability to support the bandwidth requirements of those add-on cards commonly installed into a host system. The requirements for common applications such as FDDI, SCSI, video, etc. are shown in Figure 1.1. The majority of them have fairly modest bandwidth requirements. Clearly, although a bus design can be altered to increase bandwidth, there is a performance level beyond which few applications can take advantage of such increases. Also, if the increase brings with it additional expense within the interface, its merits are of a dubious nature.

Another measure of performance is latency. Latency is the time from a master's first request to the slave's response and the completion of the access. Latency is affected by the number of other

-	Ethernet	1.25 to 1.5 MB/s
-	FDDI	12.5 MB/s
-	Token Ring	2 MB/s
-	SCSI	1 to 5 MB/s
-	Video Cards	2 to 30 MB/s
-	Co-Processor (DSP, 80x86)	5 to 20 MB/s
-	8-bit NTSC Video	6.5 MB/s
-	Laser Printers	0.2 to 5 MB/s
-	High-Quality Digital Audio	352 KB/s
-	LocalTalk/FlashTalk	30 to 100 KB/s
-	ISDN (Primary Rate)	200 KB/s
-	ISDN (Basic Rate)	16 KB/s
-	MIDI	4 KB/s

FIGURE 1.1. *Performance Requirements.*

devices waiting to use the bus and the period of time each of those devices will use the bus. The local buffering and complexity required for many applications can be reduced if the maximum latency is decreased. In addition, the performance of other applications is significantly affected by the average latency of the bus.

Ease of Use

As systems begin to perform additional functions, it is vital that they become increasingly easy to use, so that productivity remains high and frustration remains low. Within the context of an expansion bus, this implies that the add-on card must be easy to install and configure.

Small Form Factors

A primary customer concern is space on the desktop. One approach is to have the system reside on the floor next to the user's desk. A better solution, which has become more commonplace as technology has advanced, is to make the computer enclosure small enough to be placed on the desktop, or even within the monitor itself. Such small, quiet machines appeal to technical users because of their sophisticated, integrated technologies, and do not intimidate non-technical users for whom "computerphobia" is a very real productivity problem. A small form factor also allows the user to install a greater number of functions within a given volume. This significant advantage also extends to larger systems. To follow this trend, an expansion bus must have a small form factor.

Compatibility with Current Equipment

The use of any new I/O interconnect bus should not preclude the use of currently existing buses. The cost of the new bus should not be so great that it forces the system designer to choose between an existing bus and the new bus, even in larger systems. This allows the designer to augment existing buses, thereby enhancing the system, while maintaining support for older applications.

1.1.2 Pre-Existing Interfaces Can't Meet These Needs

An analysis of already existing interfaces in light of these requirements quickly shows that a good match is hard to achieve. The interfaces are either too bulky, or too slow, or too complicated, or too power-hungry (and in some cases, all of the above). The engineers who designed the SBus did so because it was the only way for them to satisfy their design requirements and goals.

A detailed comparison of the SBus and various other industry standards is the subject of the next chapter.

1.2 What is the SBus?

After the decision was made to design a new bus, it became necessary to first block it out and then design it. First strategies are chosen, and then features are determined.

1.2.1 Key Strategies

The design of the SBus stems from a few key strategies to ensure that the requirements will be met.

I/O Specialization

Most bus interfaces are general purpose. They are designed to satisfy a variety of needs, including processor-to-memory traffic, I/O transfers, and sometimes even support for multiprocessing, cache consistency, and so on. Unfortunately, such generalization often requires tradeoffs that result in an interface that is not optimal for any of these tasks.

Rather than use one general purpose interface for all the host's data transfer needs, it is possible to delegate responsibility to a variety of different interfaces. Interprocessor or processor-to-memory traffic can be handled by an MBus or Futurebus+ interface, either of which is well suited to these tasks. This allows the following SBus strategy:

Optimize SBus to provide only for the data transfer needs of I/O devices. This reduces the number of different operations that are needed, and the exception conditions that might occur. In turn, it simplifies the protocols and boosts their efficiencies.

Exploit CMOS Technologies

Current trends in the computer industry show a definite movement toward the use of CMOS ASIC technologies. CMOS has long since shed its reputation as a slow, sensitive logic family, and now offers extremely high performance, while retaining the low power levels that have always been its trademark. Gate arrays and ASICs are also coming into their own, as their densities increase to allow very complex functions in small, low-cost packages. A key SBus strategy then, is:

Optimize SBus around high-density CMOS ASICs and gate arrays. These devices provide a powerful tool for the design of modern systems.

Exploit High-Density Technologies

Corresponding increases in densities are also available via new packaging technologies. Skinny-DIP, ZIP, SIP, LCC, flat-pak, and SOIC packages are increasingly commonplace. They offer higher integration levels, because their footprints are usually smaller than traditional DIP or PGA packages. This is also because many of these packages allow both the component- and solder-side of the board to be used for mounting the packages. This can effectively double the board area available. This strategy can be summarized as:

Rely on surface-mount and other high-density packaging technologies. This allows the board outline to remain small and yet still offer the high functionality desired.

Support Autoconfiguration

To support in-the-field upgrades by non-technical people, it becomes increasingly desirable to eliminate requirements for technical knowledge from the user. Historically, a user wishing to install a new device has faced a number of serious difficulties, ranging from needing special tools to physically install the device, to moving jumpers or setting DIP switches, to reconfiguring a UNIX kernel or running complex setup programs.

Build all the tools and intelligence required for a card's installation into the card and into the system wherever possible. The ideal bus would have add-on cards that require no tools to install them and that auto-configure themselves with no input required from the user.

Scale to Both the High and Low Ends

Designing a bus that is adaptable to both large and small systems provides the greatest leverage of development efforts. Therefore, the SBus was designed with both ends of the spectrum in mind, and the following strategies were identified:

Eliminate or reduce the need for local data buffers. If latency can be kept low enough, and bandwidth high enough, the need for local buffers on an expansion card can be reduced or eliminated. This, in turn, provides savings in cost, complexity, and board or

ASIC die area, and is particularly important in low-end or mid-range machines, which are now usually desktops.

<u>Off-load the main processor as much as possible.</u> This can be done by providing facilities that allow expansion devices to contribute their own resources toward data storage and retrieval functions, control, and memory management. This generally includes DMA, interrupt, and other capabilities, and is particularly important in high-end machines, where the central processor is already faced with a large number of tasks and responsibilities.

Provide for the Bus' Commercial Success

Ensuring the commercial success of a bus requires meeting the needs of third-party vendors as well as end users. The following strategies address the vendor's concerns:

<u>Maximize opportunities for profit.</u> Business opportunity is a function of market size, growth potential, and the market share available to any given vendor. If one looks at the size and growth rate of the Unix workstation marketplace in general and that of Sun workstations (or Sun-compatibles) in particular, the size and growth potential for SBus-based machines provides a very large incentive for a vendor to bring an SBus product to market.

<u>Free the bus from</u> confidentiality and licensing entanglements. This reduces the cost of developing and distributing products.

<u>Reduce the complexity of the interface design.</u> This reduces time to market and initial investment.

<u>Provide high-quality technical support.</u> This minimizes the risk to third-parties that port applications to the SBus.

<u>Maintain architectural independence.</u> If the interface is not tied to any one processor or system architecture, then the available market and potential for its use is much larger than it would be otherwise.

<u>Work with, not against, existing interfaces.</u> This establishes a new niche, rather than competing in established territory. The new kid on the block does not want to start by making enemies. Also, existing technologies can better serve as a platform to work from than as a competitor to work against.

1.2.2 Key Features

The SBus includes a number of key features that resulted from the design requirements and the strategies discussed above.

CMOS Compatible Interface

In order to satisfy the needs for high levels of integration and low levels of power, the SBus electrical interface was designed to be directly compatible with a number of CMOS logic families. This includes CMOS ASICs and gate arrays, which can connect directly to SBus even with low-current drivers. This is advantageous, because it eliminates the need for separate bus drivers and receivers, and maximizes the number of pins available on ASICs and gate arrays that can be used for signals. (When higher currents are used, additional pins are required for grounds to reduce ground bounce.)

Virtual Memory

The SBus uses virtual, not physical addresses. This lets option cards easily share a system's memory management tools, which simplifies the programmer's task of writing I/O drivers. For example, scatter-gather operations (intended to consolidate memory usage and eliminate gaps caused by the de-allocation of space over time) aren't needed when the system allocates large contiguous blocks of memory to a device. Also, this allocated memory can easily be mapped to cover as much or as little of the available address space as is necessary. Yet another advantage is the ability to soft-map a device's base address. This eliminates the need to either set an address on the device (through DIP-switches or some other mechanism), or the need to custom-configure the driver with the device's slot-based address.

Simple, Synchronous Protocols

The SBus protocols are synchronous. This vastly simplifies the design of an SBus interface, which allows third-party vendors to reduce their time to market and improve the quality of their products. The protocols' synchronous nature also improves performance by eliminating the need to pay synchronization penalties on bus signals.

Small Form Factors

The decision to capitalize on ASICs, gate arrays, and surface mount and other high-density packaging technologies made it possible to constrain an SBus card to a very small volume, as shown in Figure 1.2. This outline and profile fit very nicely within the form factor requirements of modern desktop-type machines, while retaining the capability to support very advanced and sophisticated functions. A double-width card is also defined, providing the extra area required for particularly complex functions.

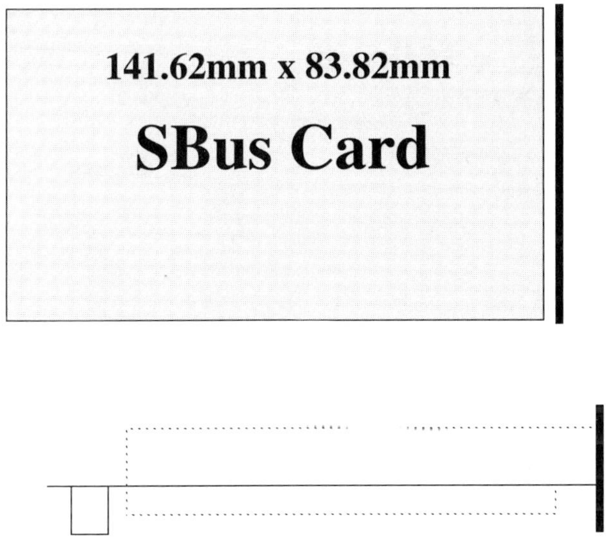

FIGURE 1.2. *Single Width SBus Card.*

The SBus card dimensions also mesh nicely with larger, standardized formats such as the 6Ux160mm EUROCARD often used in VME applications, as shown in Figure 1.3. Here, two SBus add-ons fit onto what might be a VME processor, for example. A 9U Eurocard could easily house up to four SBus add-ons. In either case, the package just fits within two standard 0.8-inch (in.) pitch card slots.

The value of such an approach is that the many advantages that SBus provides could be easily retro-fitted into an existing VME, Multibus, or other similar application.

Low Power

Substantial power savings result from optimizing SBus around highly integrated CMOS technologies. Therefore, the power dissipation on a single-width SBus card can be limited to a total of 10.7 watts (W) [2 amps (A) at 5 volts (V), and 30 milliamps (mA) for each of the 12-V supplies], without significantly restricting functionality.

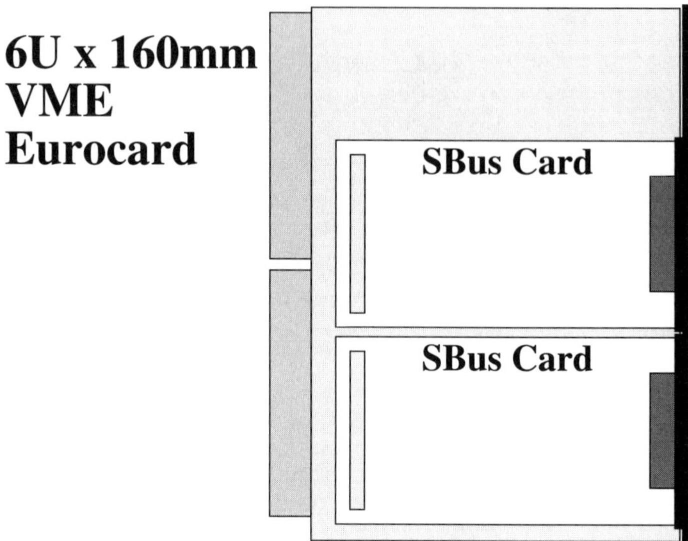

FIGURE 1.3. *Form Factor Comparison.*

A number of beneficial side-effects result from this limit. For example, power supplies can be made smaller, further reducing both the form factor and the power dissipation. Also, it is not necessary to specify airflow or other cooling requirements, and so fans can be made smaller or eliminated altogether, depending on the rest of the system's environment. This further reduces both the form factor, cost, and power dissipation.

Architectural Flexibility

Unlike many previous buses that were created by processor manufacturers, and whose signals and timing mimicked that processor architecture, the SBus is processor-independent. An SBus-based system can be implemented with equal ease and at comparable cost regardless of the processor or processors on which the host architecture is based.

High Bandwidth / Low Latency

The SBus was designed to provide sufficient bandwidth for several devices, such as those whose requirements were listed in Figure 1.1. At the same time, latency (a measure of the time that elapses between a transfer request and its completion) has been kept to a minimum. This reduces or eliminates the need for an SBus device to provide data buffers.

Open Boot

Open Boot is a mechanism that allows the host and add-on devices to easily share information about how the other operates. Also, devices can be configured, drivers can be identified and dynamically loaded, and so on. One major advantage of Open Boot's mechanisms for sharing data and configuring devices is that it provides tools to support the SBus' autoconfiguration goals.

Open Boot derives its name from what is perhaps its most important feature: the ability to use boot devices added (or even designed) after the machine is shipped. This is important, because it maximizes the system's flexibility. It also gives the system the ability to adapt to new technologies as they become available.

Open Specification and Proactive Approach

To help foster development, SBus is an open interface, without licenses, fees, or restrictions of any kind. Revision 13.0 of the specification was developed with third party feedback. As of this writing, the IEEE is sponsoring and overseeing the next revision of the specification.

1.3 Architecture

There are a number of other SBus architectural features that are very important, including:

1.3.1 Multi-Master DMA

SBus supports DMA not only to and from the host memory, but also between SBus cards themselves. DMA operations eliminate the need to first move data to memory, and then from memory to the destination device. Also, memory bandwidth is not consumed during card-to-card transfers.

1.3.2 Virtual Address Translation

Virtual memory support is an important feature which simplifies SBus development projects and aids autoconfigurability.

Development is simplified because sophisticated memory management functions can be handled within the hardware rather than the I/O driver. For example, a driver that needs a large address space simply allocates it, rather than performing the "scatter-gather" operations that might otherwise be necessary.

1.3.3 Geographic Addressing

The use of geographic addressing simplifies the design of slave interfaces, and reduces required board area.

1.3.4 Burst Transfers

The SBus provides burst transfers for the most efficient transport of data. Burst (also called "block") transfers improve the efficiency of a data transfer by distributing a single transfer's overhead across a larger amount of data.

1.3.5 Slave Flow Control

Flow control allows the slave to control the rate at which it accepts or provides data. This simplifies the design of slow slaves or slaves that occasionally become busy. Slaves handshake on each transfer in a pipelined fashion that allows them complete control over the rate of transfer without adding additional overhead to the transfer. This eases the design of SBus cards.

1.3.6 Retry Mechanism

A slave can ask that an attempted transfer be retried again later if the slave is not immediately ready. This maximizes the available bus bandwidth by not requiring the bus to wait until the slave is prepared to proceed. It also reduces software overhead because fewer interruptions of the system are required to avoid loss of data.

To support this, the slave is informed at the outset of the transfer's size. This allows a slave that temporarily cannot handle a large transfer to delay the master for a short period of time without losing data. This can reduce the buffering required on the slave.

1.3.7 Dynamic Bus Sizing

The SBus provides dynamic bus sizing, which allows the slave to control the data width it accepts during non-burst transfers. A master may transfer SBus data in 32-, 16-, or 8-bit slices. A transfer can be attempted without regard to the slave port's actual width. If a size mismatch occurs, the slave indicates this by its acknowledgment to the original transfer, and additional transfer cycles will be initiated by the master, as necessary.

1.3.8 Central Arbitration

Central arbitration provides a single place in the system to control the allocation of the bus. This is considerably cheaper than distributing this function on each add-on card.

1.3.9 Interrupts

The seven interrupt levels supported by the SBus provide the flexibility needed by multiple applications in a complex multitasking environment. This allows systems designers to manage the interrupt latencies required by different applications.

1.4 Conclusions

SBus was designed to fill a previously vacant niche in the I/O bus spectrum in a way that meets the evolving needs of an expanding user community. As a result, SBus now enjoys wide acceptance and a rapidly expanding installed base in excess of 500,000 slots. This provides unprecedented opportunity for third-party vendors, and their response has meant a broad range of choices for the computer system user. Not only has SBus fulfilled the need for performance, low cost, small size, and case of design, but it also provides substantial flexibility. This results in a faster time to market for the designer, and lower cost to the customer. SBus is a major step forward in the evolution of I/O interconnect buses.

References

Lyle, J., M. Sodos, and S. Carrie, "The SBus: Designed and Optimized for the User, the Developer, and the Manager." *BUSCON-East Conference Proceedings*, October 1990

Contrasts with Other Buses

2

In the field of natural history, any living thing is often best understood by first examining its environment. The geography and climate provide the framework in which the organism must survive. It must also share this habitat with other living things, both plant and animal, with which it either coexists or competes. As circumstances change the organism must adapt or evolve, and so to understand the present form, it is useful to look at the pressures that molded it.

The same is true when examining bus architectures. To understand the tradeoffs made in any given bus design, it is useful to understand the environment into which the bus is "born." The new bus must compete effectively against some buses and avoid competition with others. It must be adaptable enough so that it can continue to survive as the market climate changes over time. And technical and marketing pressures must be balanced against the advantages of using an existing, proven technology.

The purpose of this chapter is to briefly look at some of the other bus architectures that exist. By contrasting them with the SBus , the nature of SBus' unique design and the reasons for it will become clearer.

One very clear distinction between the SBus and all other bus architectures discussed here is in the area of virtual addressing. SBus is the only bus interface with integral virtual memory support and translation. All other interfaces use only physical addresses.

2.1 Traditional Backplane Buses

A backplane bus is one in which power and signals are transferred through a dedicated PC board that contains few, if any, active components. Prominent examples of this type of bus would be VME,

Multibus, and Futurebus+. Such buses were, and still are, suitable for use in large complex systems. Their large form factors, high power consumptions, costs, and increased complexity make them unsuitable for desktop systems and unsophisticated users. Anyone old enough to have ever tinkered with an S-100 based home computer system has first-hand knowledge of why this is so.

An example of what a typical backplane bus-based system might look like is given in Figure 2.1. The backplane is most often a row of evenly space connectors whose pins are normally daisy-chained to the corresponding pins of all other connectors. Backplanes usually also contain heavy traces or planes to distribute power to the individual boards, and must be mechanically quite stiff to retain their shape when subjected to board insertion and extraction forces. The only components you might expect to find on the backplane would be mostly passive, and used for bus termination.

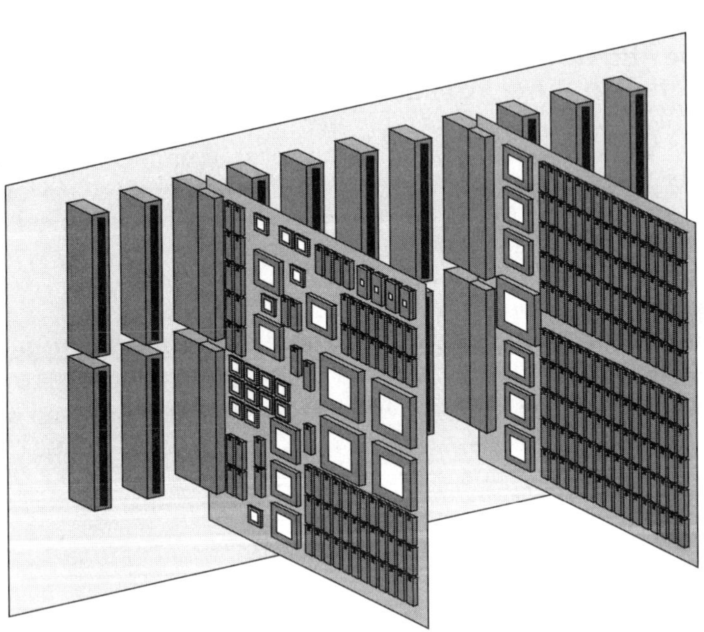

FIGURE 2.1. *Backplane Bus Structure.*

2.1.1 Futurebus+

Futurebus+ is a product of the IEEE Futurebus Working Group and the VME International Trade Association (VITA). It is a multiprocessor-oriented scalable bus architecture that is aimed at applications requiring very high bandwidths. Futurebus+ is most often found in backplane environments where the physical distances between boards are great enough that transmission line effects must be taken into account.

There are some similarities between Futurebus+ and SBus. Both can be used as chip-level interconnects, for example, directly connecting ASICs or subfunctions together. Both offer 32- and 64-bit data width options. For what Futurebus+ terms "compelled" transactions (those in which the slave is compelled to provide a response before the master proceeds to the next transfer), both buses offer approximately equivalent data rates at equivalent data widths. Both SBus and Futurebus+ support burst transfers and provide useful mechanisms for promoting a "plug-and-play" (automatic, jumperless, and switchless) approach to card installation and system autoconfiguration. Also, 82 signals are used by SBus and Futurebus+ (in the latter's 32-bit compelled subset without tagged memory support).

Although SBus and Futurebus+ share these similarities, in some ways this type of comparison is unreasonable. The intended uses of these two interfaces are so unlike each other that ultimately the differences are much more important. SBus is optimized for a chip or small module interconnect environment, and is highly specialized for its primary function as an I/O interface. Futurebus+ is more generalized, and is designed to connect small, medium, or large modules in true multiprocessor environments that require cache coherency. The differences between the two buses are reflected in a variety of ways.

One difference is in the choice of technologies used by the two buses. Although SBus was intentionally optimized for low-power CMOS environments, significant effort has been put into Futurebus+ to maintain a large degree of technology independence. CMOS, TTL, ECL, BTL, or almost any other logic family can be used (although the industry seems to be standardizing on BTL). Where SBus uses tri stateable output drivers on what are assumed to be lumped-capacitive loads, Futurebus+ uses "wired-or" drivers, which allow incident wave switching on transmission lines.

There are obvious differences in the form factors of SBus and Futurebus+, too. The SBus form factor is very small, which well suits its mission as an on-board, modular interconnect structure.

Timing and electrical drive constraints typically limit SBus to 3 or 4 slots. The currently specified Futurebus+ form factor for Profiles A and B is the much larger 12SU x 300 Eurocard format. These boards are 300 millimeters high and 300.5 mm deep, and will be capable of driving up to 20 cards. Several other profiles are possible and are being considered.

Some of these differences in form factors are shown in Figure 2.2. The top-down view shows that at least three SBus cards could fit within the Futurebus+ card. The edge-on view shows that the SBus card component heights are narrow enough that both the SBus card and its motherboard fit within a 1.6-in. space (this is equivalent to two slots in a Futurebus+ or VME card cage).

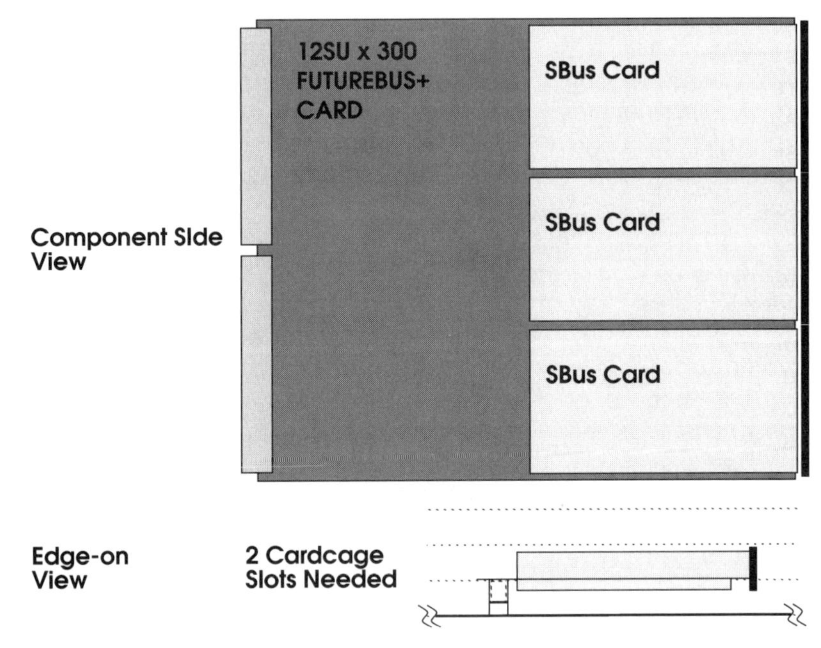

FIGURE 2.2. *Comparison of SBus and FUTUREBUS+ Form Factors.*

SBus cannot be a backplane bus. It will always be a component of a system, and there is no difficulty with respect to the placement of the required control functions (such as arbitration, support of interrupts, and timeout functions). They can always be designed into the motherboard (or equivalent). Futurebus+ developers cannot make the same assumption, however. There might be only one card in a system, or 20, but the control functions must be present and they won't be on the backplane. Some degree of control duplication was probably inevitable in such a situation, especially when multiple vendors might be involved. Futurebus+ designers have recognized this, and have distributed the control functions in such a way that the burden is minimized and the gains are maximized.

One consequence of Futurebus+'s distributed control is that there is no central clock source. Futurebus+ is also potentially much larger than the SBus, so clock skew could have been a problem. In addition, Futurebus+'s scalability goals were incompatible with specifying a fixed-frequency clock. For these and other reasons, Futurebus+ protocols are asynchronous, and center around source-synchronized beats. A beat is defined as a transition on a synchronization line by the source (a bus master) followed by the release of an acknowledge line by one or more destinations (the bus slaves). The SBus has avoided these issues and the need for such a sophisticated solution by narrowing its scope. The use of a centrally-generated clock of fixed frequency greatly simplifies SBus protocols and is a major difference between SBus and Futurebus+.

Unlike SBus, Futurebus+ is designed for systems where copies of data can be kept in multiple locations (caches) simultaneously, and where all such copies must remain in sync with each other. Cache coherency protocols, which SBus does not need, are an integral part of Futurebus+. These protocols allow multiprocessors to share data efficiently and accurately, while retaining the performance benefits that local caches afford them.

In addition to the cache coherency mechanisms, Futurebus+ also offers a greater selection of transfer modes than does SBus, in order to suit its more general purpose multiprocessing role. Among these are broadcast transfers, which move data between one master and multiple slaves simultaneously. Split transactions can also be used, in which a transfer of long or unknown duration can be initiated and concluded in two separate phases.

Despite some similarities, Futurebus+ and SBus are very different. This is a reflection of the different problems they are meant to solve. The two do not compete, but in fact are compatible and can be used symbiotically. Futurebus+ could be used as a very high bandwidth interface between one or more processors and system memory. SBus could be the path through which I/O options are customized easily and at low cost. There are many benefits to such a symbiotic approach, and companies such as Sun Microsystems and Motorola have publicly announced support for both SBus and Futurebus+.

2.1.2 VMEbus

VMEbus is derived from Motorola's VERSAbus. Motorola's Munich facility adapted this interface standard to the standard, more compact Eurocard format, calling the hybrid VERSAmodule Eurocard; hence the "VME" acronym. This combination of protocols and form factors quickly became a standard in Europe. In 1981, Motorola, Signetics/Phillips, Mostek, and Thomson CSF released the VMEbus into the public domain and announced that it was a nonproprietary standard. VMEbus is now enormously popular, and extensions such as the recent VME64 promise to keep it alive well into the future.

Many comparisons between VMEbus and SBus are similar to those made in the previous section regarding Futurebus+, and so only a few additional comments will be made here.

Interestingly, Sun Microsystems developed the SBus specifically to replace the VMEbus in desktop workstations manufactured by the company. VMEbus systems can include up to 21 slots. As the technology is bipolar TTL-based, the signal loading can be quite high. Also, the bus is a transmission line with a low impedance, and so low-value terminations are found at both ends of most signal lines. For both of these reasons, very heavy drivers, which consume a lot of power, are required. As a backplane bus, it is also physically large. (See Figure 2.3.) Both of these factors place ever increasing burdens on Sun's efforts to build smaller, lighter, quieter, and less expensive machines.

Because VMEbus is a backplane bus, there must be some provision for bus-controller functions to reside on at least one card in the system. In the case of the VMEbus, this includes the clock generator, the bus timer, the arbiter, and the interrupt acknowledge daisy-chain driver (more on this later). For simplicity's sake, the decision was made to require these functions to be in Slot 1. Unfortunately, any card that is the only card in a system, or contains the only master in a system otherwise composed only of slaves (a slave-only card cannot be the only card in a system), must contain these control functions. These are disabled if the card is plugged into any slot other than Slot 1. Usually, this is done by means of a jumper on the board. This is a simple approach, but leads to wasted logic in many configurations and is not conducive to achieving the automatic configuration goals for Sun's newer machines. This is another reason a replacement for VME was needed.

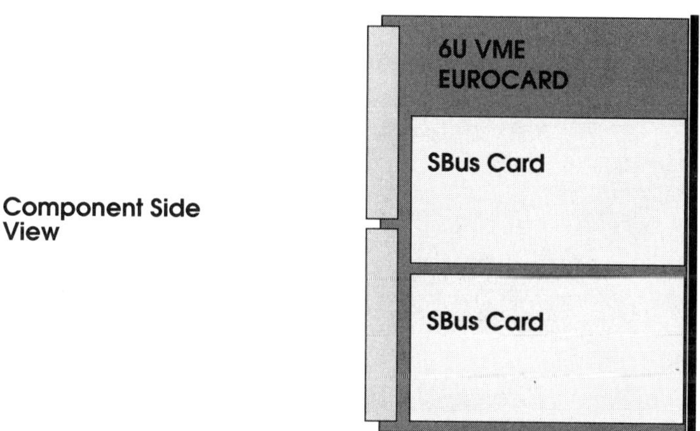

FIGURE 2.3. *Comparison of SBus and VME Form Factors.*

Sun has not abandoned the VMEbus, however, because VMEbus' usefulness in deskside and server applications will continue for some time. Also, like SBus and Futurebus+, SBus and VMEbus can work together well in some applications, such as embedded processing or control. The SPARCengine 2E uses both VME and SBus interfaces, and helps to bring the advantages of SBus into existing VMEbus environments. This is shown in Figure 2.4.

2.2 Daughter-card Buses

Daughter-card buses take a more integrated approach than do backplane buses. Standard elements that every system must have are built into a single motherboard. This generally includes the CPU, system memory, the bus controller, and all associated support circuitry. Options such as additional serial ports, extra memory, or video graphics adapters are built on smaller daughter boards that connect to the motherboard, and can be mixed and matched to suit the end user's needs.

FIGURE 2.4. *SPARCengine 2E Combines both SBus and VME Interfaces.*

The structure of what a typical daughter-card-based system might look like is shown in Figure 2.5. In this case, three out of five available slots are filled with daughter-cards. These boards are perpendicular to the larger motherboard.

These buses are usually smaller, draw less power, and cost less than backplane buses. Common examples of daughter-card buses are the AT bus, EISA, Micro Channel, and NuBus.

2.2.1 Micro Channel

Micro Channel is a product of International Business Machines Corporation (IBM). It is the I/O expansion mechanism that is used

in IBM's Personal Systems/2 family of personal computers, and in
the RS/6000 family of RISC-based workstations.

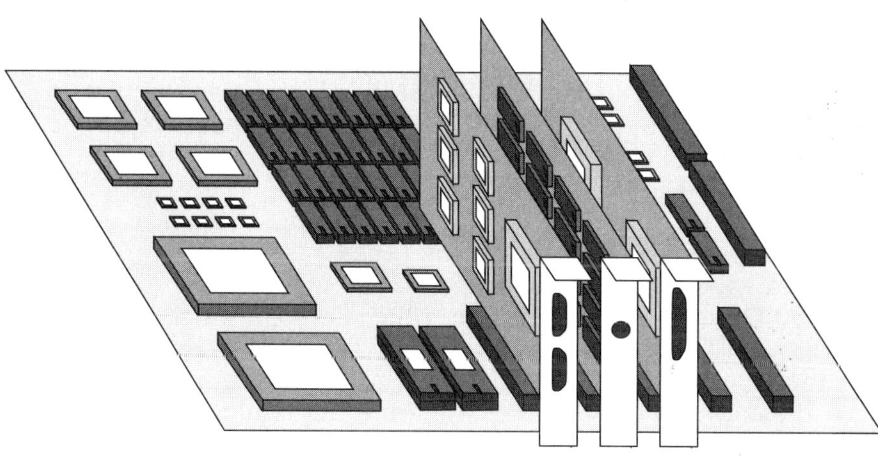

FIGURE 2.5. *Daughter-Card Bus Structure.*

Micro Channel and SBus share a number of similarities. Both are
aimed squarely at I/O expansion with multiple DMA masters, in
highly integrated environments that rely extensively on ASIC technology. Both have burst modes, bus sizing, slave flow control,
shareable interrupts, and centralized arbitration and control.
Micro Channel and SBus allow similar amounts of power (12.6 W
total and 10.7 W total respectively, 2 A at 5 V in each case). Both
Micro Channel and SBus also have 32-bit standard data widths,
with extensions that allow 64-bit transfers.

Micro Channel offers 32-bit physical addressing, and allows a
24-bit subset. There is also a 16-bit I/O address space that is completely distinct from the regular memory space. These addresses
must be shared among up to eight Micro Channel slots. SBus, on
the other hand, offers 32-bit virtual addresses and 28-bit physical
addresses (some machines only support a 25-bit subset). A key distinction, though, is that SBus physical addresses are *per slot*; each
device has its own address space and does not need to share physical addresses. SBus provides a geographic select to each slot,
informing that card when to participate in a bus transfer, without

the need for the add-on card to perform any pre-decodes on the address. This is an advantage made possible by SBus' virtual addressing capabilities, and the MMU function contained within SBus controllers.

Both SBus and Micro Channel are symmetric architectures, in that any DMA master can communicate with any slave. Unlike the SBus, however, Micro Channel is dependent on processor architecture. It is very closely tied to Intel's processors (8086/80286/80386/ etc.). For example, the 16-bit I/O address space mentioned in the previous paragraph is accessed through a special series of operations in these processors' instruction sets. The timings and signal names, too, are closely related to the bus interface definitions of these processors. The SBus is processor independent, and can be used with any kind of RISC or CISC processor.

One very unique feature that Micro Channel has is its built-in support for audio and video applications. The bus' designers have included a signal that can be used to perform analog sums of audio signals, with the result being amplified and used to drive the system's speaker. There is also a system-dependent Auxiliary Video Extension, which can allow video resources to be shared. These mechanisms are an acknowledgment of the multi-media applications that might become commonplace in the kinds of environments in which Micro Channel is used.

Standard Micro Channel transfers require 200 nanoseconds (ns) for every 4 bytes of data. This results in a bandwidth of 20 megabytes (MB)/second (s). This contrasts sharply with SBus' 80 MB/s sustainable rate. Micro Channel does offer "Matched Memory" options, which might allow faster transfers in some cases, but these are system dependent, so achievable performance figures are unclear. Micro Channel B is a higher performance variant of Micro Channel, which claims transfer rates of up to 160 MB/s at 64-bit widths, equivalent to SBus' rate at these widths.

Although SBus is optimized for I/O purposes, Micro Channel is more general-purpose. System memory and I/O share the bus in this case, which has two interesting effects on performance. The first is that Micro Channel memory accesses subtract from the bandwidth that would ordinarily be available for I/O operations (and vice versa). The other is that the Micro Channel Specification defines refresh operations required for dynamic RAM, which also subtracts from available bandwidths.

Both Micro Channel and SBus also limit bus capacitance sharply in order to minimize propagation delays. In both cases, add-on cards are allowed to present no more than 20 picofarads (pF) of additional capacitance to the bus. Micro Channel limits total capacitance to 240 pF for most signals, whereas SBus limits the total to 160 pF. Micro Channel's figure is larger because it is aimed at up to eight slots, and assumes drivers that can sink at least 24 mA. Such potent drivers can be a very big problem for the high density CMOS ASICs that SBus is optimized for, however. Therefore, SBus is even more stringent in its capacitance limits and will generally only contain three or four slots.

A similar argument applies to DC leakage currents, although here the differences are even more startling. Micro Channel allows up to 1.6 mA per signal per channel connector in most cases. SBus allows only 30 microamps (µA). This is a difference of about 50 to 1! Because DC loading is also very critical to high-density ASICs, SBus' lower power interface will allow higher levels of integration to be achieved.

The Micro Channel and SBus form factors are compared in Figure 2.6. The Micro Channel card's major dimensions are 3.476 in. by 11.500 in. This is significantly larger than an SBus card (whose outline is superimposed). SBus is substantially smaller, which makes it the more attractive add-on for desktop or laptop machines.

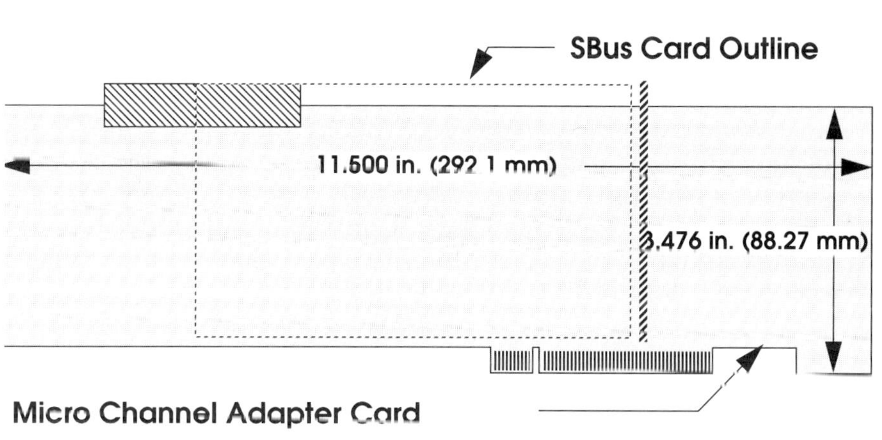

FIGURE 2.6. *Comparison of SBus and Micro Channel Form Factors.*

Micro Channel does use geographical addressing in a very limited way. Micro Channel's geographic addressing mechanism is used only for the Programmable Option Select (POS) registers, though, and not for more general accesses to the card. These registers are used to help automate the configuration process, much like SBus uses an ID PROM. In both cases, these elements contain information about what the card is and what it is capable of. They also provide a mechanism for enabling or disabling certain features, configuring addresses, and so on.

Micro Channel's documentation calls it a synchronous interface, and there are asynchronous extensions that allow slaves with access times greater than or equal to 300 nanoseconds to be accessed. It is important to recognize, though, that IBM's use of the terms synchronous and asynchronous differs substantially from the way they are defined in this book. Micro Channel does not use any kind of bus clock to derive timing or sequence information within its protocols. Most Micro Channel timing diagrams do not even show a clock. There is an "OSC" signal, which is 14.318 megahertz (MHz), but it is intended to be used in deriving video dot clocks, baud rates, etc. It is not necessarily synchronized to the bus activity in any way.

IBM's use of the term synchronous is used to imply that the starting and end points of a transfer are well-defined, and are related (or synchronized) explicitly to transitions on control lines generated by the master device. Futurebus+ uses the term source-synchronized in a similar way. The term asynchronous is used within the Micro Channel literature with respect to extended cycles whose actual length is controlled by the slave, not the master.

When SBus is called synchronous, however, there is a completely different definition in mind, and one that seems more consistent with the industry norm. All signal timings are derived explicitly from the SBus Clock. Control signals must be qualified and sampled with respect to this clock. Signals must also meet setup and hold requirements with respect to the clock, and state transitions occur at clock edges. Using this definition, Micro Channel is completely asynchronous, and its protocol even asks for unlatched decodes in many cases. This difference is significant, because synchronous logic and state machines are much easier to design and verify than asynchronous logic.

One very substantial difference between these two buses is that whereas SBus is open and available free of charge to anyone that wishes to use it, Micro Channel is proprietary. It even sounds proprietary. DMA cycles initiated by option cards are called Third-Party cycles. More importantly, though, stiff license fees must be paid by products that are developed for Micro Channel. These fees include a percentage of revenues acquired, and a retroactive fee for any past use of the AT interface. This adds cost to any end-product, and it also is a sore point for manufacturers, who have balked in some cases and are now backing the rival EISA architecture. This could be a major factor behind Micro Channel's relatively slow growth, and ultimately might lead to its demise. Even IBM seems to be throwing in the towel to some extent, because several recent personal computer products (including the PS/2 models 35LS, 35SX, and 40SX) have chosen the AT bus instead of Micro Channel.

In summary, SBus and Micro Channel share many features and concepts. SBus, however, offers higher performance and integration levels. Ultimately, it might also offer lower costs and an open ended market, because SBus is not burdened with royalties.

2.2.2 ISA and EISA

The Industry Standard Architecture (ISA) is a formalization of the bus interface definition used by IBM PC AT and compatible personal computers. This was itself an evolution of a bus design dating all the way back to the IBM PC and compatibles. The Extended Industry Standard Architecture (EISA) extends this definition, adding capabilities, while retaining backwards compatibility. Both are products of a consortium of PC clone manufacturers, who were not prepared to follow IBM's transition to Micro Channel (and to pay royalties and license fees as a result). Often called the "Gang of Nine," because there were originally nine member companies, this consortium sought to provide the personal computer industry with a high performance bus alternative that was not proprietary, and that did not render obsolete the myriad of systems and add-ons already on the market.

ISA and EISA interfaces are both helped and hindered by their installed base of PC-, XT-, and AT-compatible machines and cards. Any host built around either of these interfaces benefits from an immense, diverse selection of add-on cards already available. This is a problem, though, when limitations in the previous products cannot be fully overcome. For example, ISA interrupts cannot be shared easily because they are edge-triggered. EISA systems can share interrupts, but only if all devices on any interrupt level are EISA compatible. Even one ISA-only interrupt driver prevents all others from sharing that interrupt. Also, like Micro Channel and SBus, EISA contains an autoconfigurability mechanism. ISA cards do not, however, and will still need to be manually configured in EISA machines. Problems such as these are likely to be common for a long time, too. Because ISA cards will work with EISA machines, there is often little incentive for developers to update their designs. For this reason, too, it will probably be quite some time before the 32-bit data width of EISA can be greatly utilized; most existing cards are only 8 or 16 bits wide (the ISA maximum). All this has given ISA/EISA a leg up into the marketplace, but one that ultimately might prove more of a hindrance than a help.

Because ISA, EISA, and Micro Channel all share common ancestors, they are more alike than different. EISA is a bit faster than Micro Channel (33 MB/s vs. 20 MB/s respectively; 8 MB/s for ISA), and can support more masters (15 for EISA, 8 for Micro Channel). An extensive comparison with SBus is unnecessary here, because the critical distinctions were drawn in the Micro Channel section. Only a few points warrant a further look.

When compared to SBus, the differences found in power and form factor are most obvious. ISA/EISA are very power-hungry, consuming up to 4.5 A at 5 V, 1.5 A at +12 V, 300 mA at −12 V, and 200 mA at −5 V (a voltage not supplied by either Micro Channel or SBus). At just over 45 W total, that's over 4 times greater than SBus' 10.7 W. The dimensions of an ISA/EISA add-on card are 13.13 in. by 4.5 in., giving it just over 3 times the area of an SBus card. Both of these factors make SBus a much more attractive choice for highly integrated or portable applications.

2.2.3 NuBus

NuBus is actually kind of a missing link between backplane buses and daughter-card buses. Unlike Micro Channel, ISA, EISA, and SBus, NuBus has no architecturally distinct bus controller (except for the clock generator). Like Futurebus+, all NuBus devices are peers and all participate in system control functions such as arbitration. There is even a triple-height Eurocard form factor that is most suitable for backplane environments. Despite these backplane tendencies, however, most applications of NuBus now use the desktop form factor (compared to SBus' in Figure 2.7.), with a structure that at least physically resembles that found in Figure 2.5. This is why NuBus is still considered a daughter-card bus for the purposes of this discussion.

In many ways, NuBus was ahead of its time when its foundations were laid at MIT in the late 1970s. Many of its key goals and requirements foreshadowed those of SBus, and there are many similarities in the two buses. In both buses, the protocols are very simple and are synchronous to a bus clock. The protocols of each are independent of processor type, and the overall architectures allow multiple masters and slaves to communicate symmetrically. Transfers are 8, 16, or 32 bits wide, optionally with parity protection, and block transfers are supported.

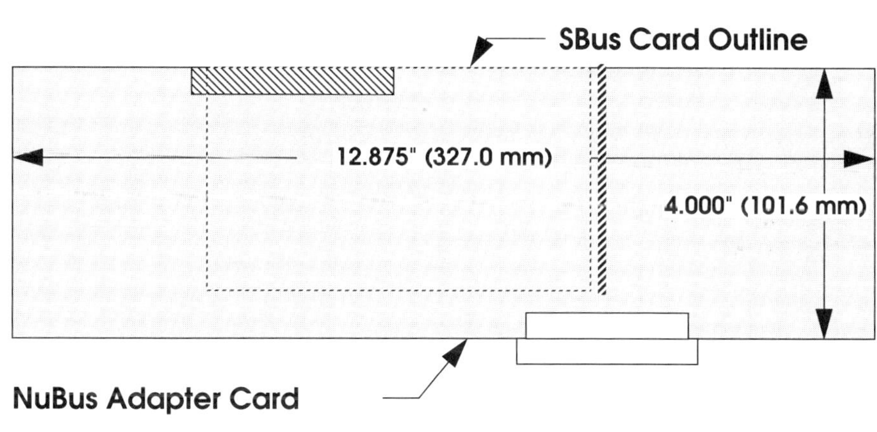

FIGURE 2.7. *Comparison of SBus and NuBus Form Factors.*

NuBus multiplexes 32-bit physical addresses with the data lines (SBus multiplexes 32-bit virtual addresses with the data lines, but provides dedicated physical address lines). This address space is shared by all slots, except for a small portion that is dedicated to slot space. Each of up to 15 NuBus slots has wired into it a unique 4-bit ID field, which is used to qualify part of any address in the slot space. The result is that each slot has a small, dedicated address space to which it alone can respond. This is a form of geographic addressing, and like SBus' geographic select signals, it is useful for autoconfiguration purposes.

Fifteen available slots allow for great flexibility in configuration, but there are problems associated with them as well. Bus latencies can be very long. On NuBus, 15 masters x 25.5 microseconds/transaction = 383 μs, or longer, if multiple locked transactions are allowed to occur. This is a long time to wait if you have a fast

I/O device without much local buffering. A 4-master SBus system at 25 MHz bus clock rates requires less than 8 µs.

NuBus' age is a detriment in some ways, now, because of advances in technology made since it was defined. The clock rate is set at 10 MHz, which limits the ultimate transfer rate to 20 MB/s for word transfers, 37.5 MB/s for burst transfers. Compare this with SBus, which has clock rates up to 25 MHz, and burst transfer rates of 80 MB/s. (The IEEE and Apple Computer are working on an enhanced version of the NuBus definition which will be known as P1196-R, or NuBus '90. Among the chief enhancements that will be included in this new definition are cache coherency protocols, and double clocking (in burst transfers only), which will effectively double the burst transfer bandwidth.)

NuBus is not specialized for I/O purposes the way SBus is. It is a general-purpose bus that incorporates many advanced features similar to those now found in Futurebus+. For example, bus parking, resource locking, and broadcast capabilities (called attention cycles) are included in NuBus' protocols. Also like Futurebus+, interrupts are called events and are transactions (writes into dedicated address ranges in this case), not signals. This approach is nicely shareable and symmetrical—any or all devices could interrupt any other. Basic NuBus protocols do not contain built-in provisions for cache coherency (though new revisions will), but the "snooping" and "snarfing" this would require can be accomplished using the building blocks that are already defined. All of this makes it possible for NuBus to support either loosely-coupled or tightly-coupled multiprocessing systems.

Since its inception at MIT, NuBus has had an interesting history. Western Digital built an early personal UNIX workstation around it called the Nu Machine. Texas Instruments then acquired this project and built several LISP- and UNIX-based systems around it, including the Explorer LISP machine and the System 1500 UNIX multiprocessor. NuBus remained mostly within the academic and R&D realms, though, until 1987. In that year, the IEEE Standards Board formalized NuBus' definition as IEEE Standard 1196, and Apple Computers, Inc. adopted it for use in their Macintosh II line of computers.

There is an interesting conclusion that can be drawn from this that relates back directly to one of SBus' key requirements. NuBus is a very advanced technology that is now a major name in the bus industry. For almost a decade, though, it was no more than a footnote in the bus industry, because only a small number of very specialized machines used it and it was not open; MIT licensed it to

Texas Instruments and then later authorized them to sublicense it. There was little incentive for third party developers to produce products based on it, because the profit potential was small. Also, any profit that was made was likely to be eaten up by royalties, assuming that a license could even have been obtained. Now though, its use in a high-volume commercial product has brought it off the shelf and into contention as one of the major players in the bus marketplace.

All this underscores the importance placed on both the technical and the commercial requirements when the SBus was designed. Without both, timely success and the phenomenal growth SBus has seen in such a short time could not have been achieved.

2.3 Mezzanine Buses

Mezzanine buses are similar to daughter-card buses, in that the system contains a motherboard on which the standard functions are found, and smaller modules for the options. In this case, though, the option cards are mounted parallel to the motherboard in a manner reminiscent of a mezzanine or balcony (hence the term). This affords an even greater reduction in form factors.

An example of what a mezzanine bus-based system looks like is shown in Figure 2.8. Advancing technology trends have increased the level of integration possible, while reducing power requirements. This, in turn, allows an engineer to pack ever more complex functions into smaller spaces, without undue concern over cooling issues.

SBus is a mezzanine bus. Other examples are MBus and TURBOchannel, which will both be briefly profiled in this section.

2.3.1 MBus

Contrasting SBus and MBus is very much an "apples to oranges" comparison. These buses were both designed by Sun Microsystems with the very explicit intention of making them cooperative, not competitive. SBus is a high-performance I/O interconnect, and MBus is a high-performance CPU and memory interconnect with built-in multiprocessor support. The two will often be found working together in the same system, each occupying its own specialized niche.

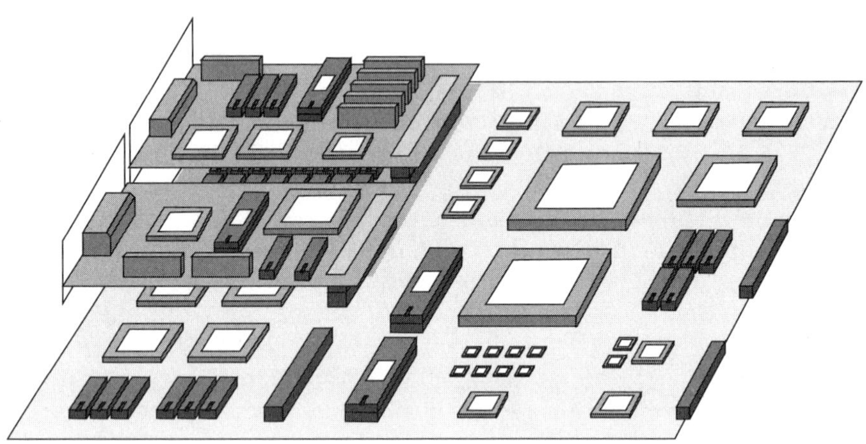

FIGURE 2.8. *Mezzanine-Card Bus Structure.*

Both MBus and SBus were designed for high-speed data transfers between circuit modules, using only a few simple transactions and a small number of signals. Both are fully synchronous, support burst transfers, and can support multiple masters.

Unlike the SBus, though, MBus is not optimized strictly for CMOS applications; it can be used with both CMOS and BI-CMOS technologies. Also, the MBus can operate at rates up to 40 MHz (compared to the 25-MHz SBus maximum), and data widths are standardized at 64 bits, whereas such a path width is only an extension of the SBus.

The key difference, of course, is in the applications targeted. SBus is an I/O bus. MBus, however, is optimized as a processor-to-memory path. MBus has two levels of compliance. Level 1 targets uniprocessor applications. Level 2 adds transactions specifically designed to support cache-coherency in multiprocessor applications.

2.3.2 TURBOchannel

Digital Equipment Corporation has designed its own mezzanine architecture, which they named TURBOchannel. This interface first went public in April of 1990 with the introduction of the DEC-station 5000 series, and like the SBus, the specification has been made open to vendors that wish to develop applications based around it.

SBus and TURBOchannel each evolved along similar lines, exploiting the same technologies and facing similar design goals: high performance, small form factors, CMOS compatibility, simple protocols, and low cost. It's not surprising, then, that the results are similar, too. For instance, both are CMOS-based synchronous interfaces that have a 25 MHz maximum clock rate and a 32-bit wide data path in the standard configuration. Also, both are primarily intended for I/O expansion applications, offer small form factors, and provide high DMA bandwidths, which now peak at 100 MB/s.

As open specifications, both SBus and TURBOchannel are designed to actively encourage third-party manufacturers to develop products for or incorporate these buses into their own systems. The reasons for this openness and support reflect a common and simple strategy; both Digital Equipment Corporation and Sun hope to concentrate their development efforts on the more generic host systems, while third parties and OEMs assume the responsibility of providing the niche products and services.

But there are important differences, as well. One area that has received perhaps the most attention is performance. The peak data rates at different burst transfer sizes for SBus and TURBOchannel are shown in Figure 2.9. The data represent sustainable bandwidths for write transfers; read transfers require at least one additional cycle for both SBus and TURBOchannel. TURBOchannel offers slightly higher bandwidths than SBus at 32-bit widths. This is because TURBOchannel requires only a two clock-cycle overhead for each data word or multiple word burst, while SBus requires four.

This streamlining has its down side, however. TURBOchannel cannot offer the advantages of direct virtual memory access. Nor can TURBOchannel offer flow control of any kind on DMA burst transfers; there is no provision for devices that might need to momentarily slow down or pause (to refresh memory or to access a new page in page-mode memories, for example). Compounding the problem, there is no up-front indication of how long a DMA transfer will take, making it impossible for a slave to "plan ahead." One consequence of this is that TURBOchannel DMA accesses are not allowed to cross 2048-byte address (or page) boundaries.

Interestingly, a TURBOchannel-based host is incapable of directly initiating DMA transactions of any sort (it is limited to programmed I/O only). Conversely, TURBOchannel options can only perform DMA operations to the host memory. Because of this, and because neither the processor nor the system memory can be placed on a TURBOchannel option, TURBOchannel is an inherently asymmetric architecture. One result is that it is not possible for any one option to communicate directly with any other. If a block of data must be moved from one TURBOchannel option to another, it must first be moved into the system's main memory with a DMA operation, and then moved with yet another DMA operation (or even worse, with programmed I/O) into the other option.

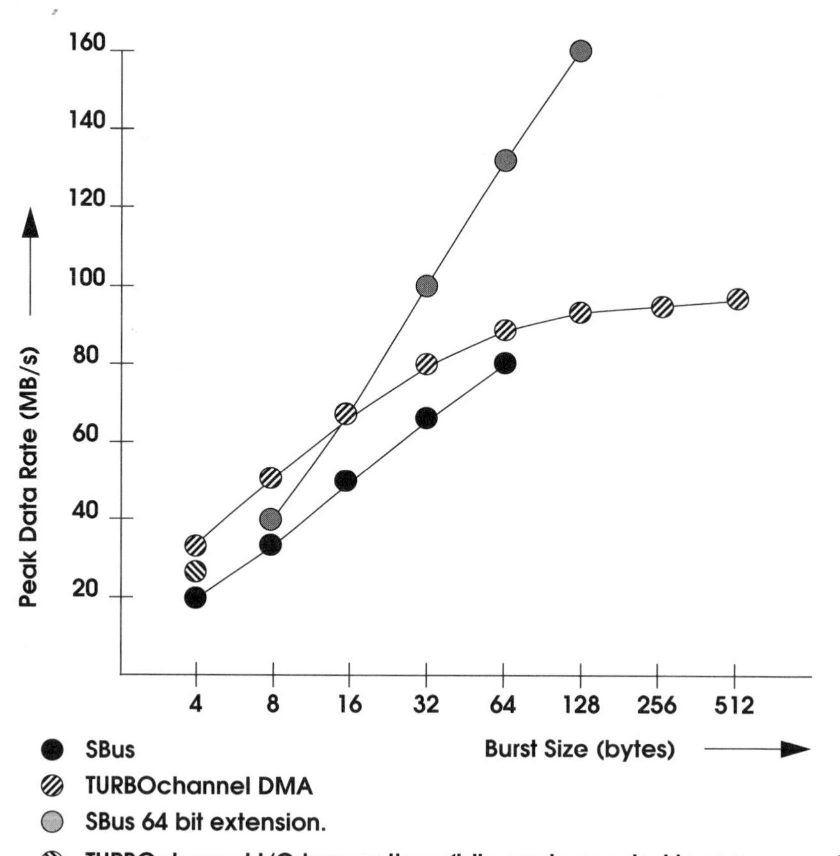

FIGURE 2.9. *Graph of Burst Size vs. Peak Data Rate (25 MHz Writes).*

The SBus architecture, on the other hand, can be used either symmetrically or asymmetrically. It is possible to build SBus systems with the processor and the system memory directly on an SBus card. Also, any DVMA master can communicate directly with any slave, and as a result, the previously discussed block move could be done directly in a single DVMA operation.

A single-width TURBOchannel option card is 4.6 in. wide by 5.675 in. long for a total of 26.1 square (sq.) in. of board area. (See Figure 2.10.) A single-width SBus card offers a total of 19.1 sq. in. (Figure 3.30 on page 100). Double-width options can be built for both buses; TURBOchannel also offers a triple-width option.

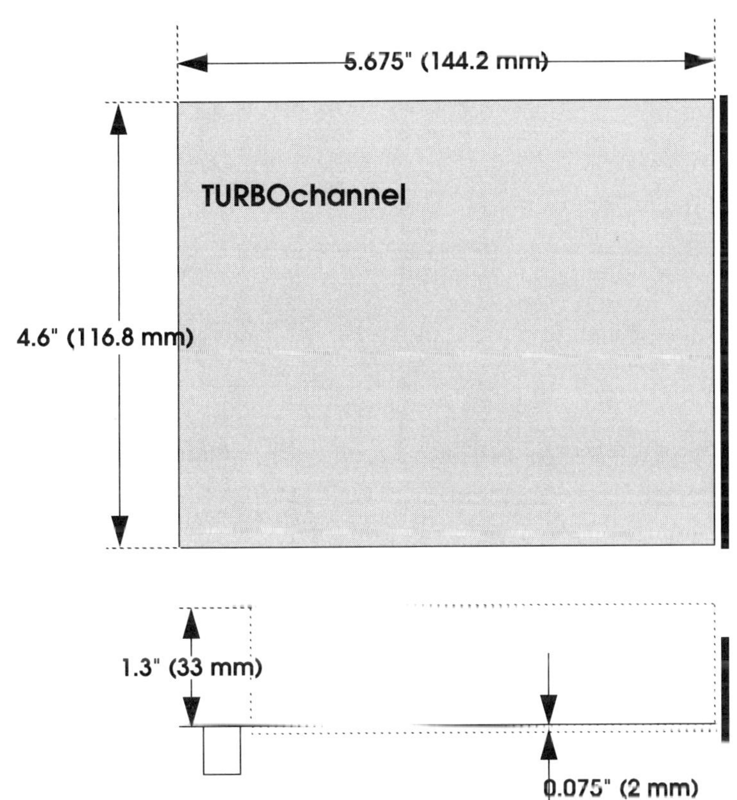

FIGURE 2.10. *TURBOchannel Single-Width Card Dimensions.*

The wider design for TURBOchannel might also restrict its use as a mezzanine bus in backplane-based applications, such as VME and FUTUREBUS+. Only one TURBOchannel option will fit without overlap onto a standard 6U-sized VMEbus Eurocard, whereas two SBus cards fit nicely. (See Figure 2.11.) The generous component height allowances specified by TURBOchannel for the component side of the board (1.3 in.; 33 mm) also means that at least three standard-sized (0.8-in. width) VME slots will be required for any such application. SBus-based applications require only two slots.

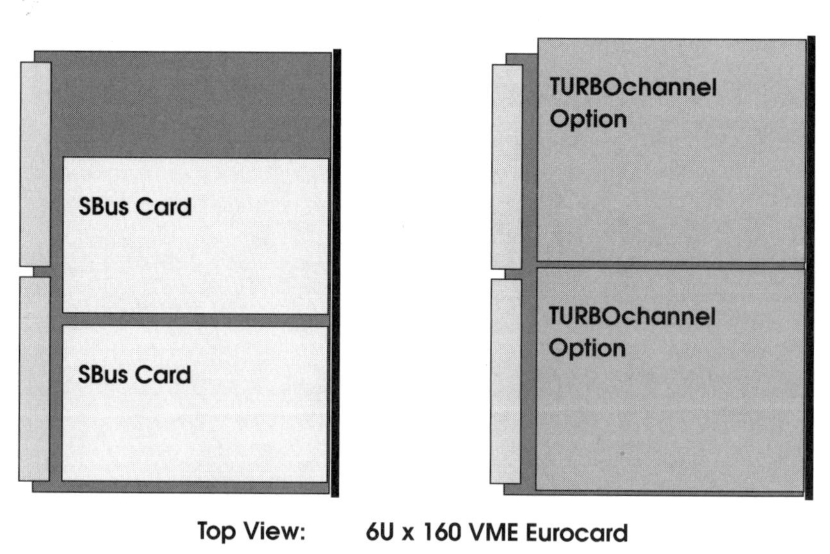

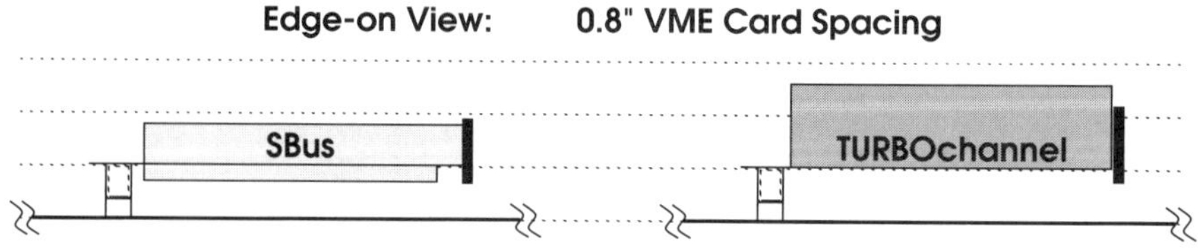

FIGURE 2.11. *SBus and TURBOchannel Comparisons to a 6U Eurocard Form Factor.*

TURBOchannel provides 4 A at 5 V and 0.5 A at 12 V to each expansion slot, or about twice what SBus allows. The power difference between specifications is a result of design trade-offs. SBus is aimed at more highly integrated applications. Limiting the SBus to approximately 11 W minimizes its required power dissipation for supply (or battery) capacity, connector power and ground pins, and system cooling. Digital Equipment Corporation had no such goals in mind for TURBOchannel, and wanted an interface that requires higher-powered drivers, and so has chosen to be less restrictive. One result, though, is that a fan of some kind is required (a minimum airflow is specified), further reducing TURBOchannel's desirability for very high integration or low power applications.

Though a number of differences between SBus and TURBOchannel exist, those differences are relatively minor when compared to their similarities. More than anything else, the existence of both SBus and TURBOchannel and the similarities between them validates the emerging need for this type of mezzanine bus.

Acknowledgments

Apple is a registered trademark and Macintosh is a registered trademark of Apple Computer, Inc.

DECstation and TURBOchannel are registered trademarks of Digital Equipment Corporation.

IBM, Micro Channel, Personal Systems/2, PS/2, and RS/6000 are registered trademarks of the International Business Machines Corporation.

Intel is a registered trademark of Intel Corporation.

NeXT is a trademark of NeXT Computer, Inc.

NuBus, Explorer, Explorer II, and Nu Machine are trademarks of Texas Instruments Incorporated.

Sun and the Sun logo are registered trademarks of Sun Microsystems, Inc. SPARCserver, SPARCstation, SPARCengine, and SunOS are trademarks of Sun Microsystems Inc.

VERSAbus is a trademark of Motorola, Inc.

References

Di Giacomo, J. ed. *Digital Bus Handbook*. McGraw Hill, 1990.

Digital Equipment Corporation. "DECstation and DECsystem 5000 Model 200 Technical Overview."

Digital Equipment Corporation. "TURBOchannel Developer's Kit - Version 2."

Digital Equipment Corporation. "TURBOchannel Overview." April 1990.

Glass, L. B. "Inside EISA." *BYTE*, November 1989.

Honan, P. "EISA under the Glass." *Personal Computing*, August 1989.

IBM Corporation. *IBM Micro Channel Supplement for the PS/2 Hardware Interface Technical Reference*. November 1989.

IEEE. IEEE P896.1 (Futurebus+ Logical Layer Specifications). Draft 8.1a.

IEEE. IEEE P896.2 (Futurebus+ Physical Layer and Profiles). Draft 4.0.

IEEE. IEEE P1101.2 (Mechanical Core Specifications for Microcomputers Using 2 mm Pitch Connectors). Revision 0.

Lewis, P.H. "IBM turns back the clock." *New York Times*. (appeared in the *San Jose Mercury News. Computing Section*. June 16, 1991).

Sun Microsystems, Inc. *SBulletin*, June 1990

Sun Microsystems, Inc. *SBus Specification, Revision B.0*, January 1991

Hardware Concepts

3.1 Fundamental Concepts

Perhaps the most fundamental concept of the SBus is that it is designed to be a chip-level I/O bus. Its purpose is to allow high-density CMOS integrated circuits to be connected with one another easily, efficiently, and with cost and performance optimized toward I/O data transfer needs. Most of its other features are ultimately derived from this concept, which is also SBus' primary distinction from other bus architectures.

One major consequence of this is the necessity to eliminate circuit complexity wherever possible. A complex bus protocol is undesirable in a chip-level bus, not just because it makes the design task more difficult, but also because it leaves less logic available for a device's primary function.

Another major consequence is the need to eliminate signal buffering. ASIC drive levels are often very low, and special assumptions and allowances must be made if the interface signals are to be driven straight from the component's pins. To do this without adding complexity (and in fact while reducing the complexity in other ways) posed a special challenge to SBus' designers.

Buffers are typically found in low- or medium-scale integration packages, and it is not unusual for several parts to be necessary to buffer the signals from a single ASIC. The number of devices in the design, the number of interconnections between these devices, the board area required, and the number of pages required in the schematics and in the documentation all increase as a result. Also, the added components increase power consumption. All of these factors negatively impact the cost, the time to market, and the chances of making a mistake or settling for a less than optimal design.

Performance can be degraded, too. Buffers add delay to a signal. In a more traditional bus environment, where both drivers and receivers are often required, there can be up to four buffers in any round-trip signal path. An example of this situation is shown in Figure 3.1. Here, some logic in Block A generates a signal, which passes through its output driver (1). Block B passes the signal through its receiver (2), and then performs some operation on it. When a response is ready at some later moment, it must then pass through B's driver (3) and A's receiver (4), before the logic in Block A can recognize it. Even with very fast buffers, the resulting extra delay can be a large portion of the overall round-trip time.

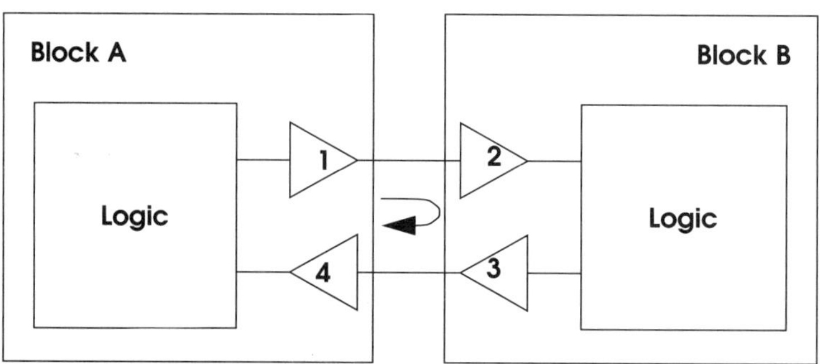

FIGURE 3.1. *Up to Four Buffers Might Be Necessary in any Round-Trip Signal Path.*

The delays associated with signal buffering also complicate the timing analysis of the overall design. Buffer delay must be accounted for when calculating signal delays, setup times, and hold times. Buffer delays are not constant, though; there are both maximum and minimum delays specified for each device, and the actual delay can vary anywhere between these two extremes. The result is that each buffer adds yet one more variable to the timing equations that must be solved, and one more chance to make a timing mistake that will result in an inoperative (or even worse, an unstable) design.

This section discusses other related fundamental concepts selected by the designers of the SBus.

3.1.1 Synchronous Sampling

A synchronous circuit is one in which all operations and changes of state occur simultaneously, usually as a result of a beat or pulse supplied by the circuit's clock. Like a metronome, this clock beats at a fixed frequency, and its purpose is to regulate the flow of the circuit. One advantage of synchronous circuits is that signal paths are relatively easily isolated, and the calculation of their delay straightforward. If this circuit design is done properly, another advantage is that signals will change state only when it is safe to do so, between beats. Setup and hold times can be guaranteed, and metastability need not be considered. This also means that the exact value of signals between beats is unimportant, and glitches (on any signal but the clock) will not affect proper circuit operation. This provides added noise immunity and limits sensitivity to factors such as ground bounce and signal ringing. Perhaps the biggest advantage of synchronous circuits, though, is that they are easily understood, designed, observed, and debugged: their state machines march forward in a logical and precise fashion.

Synchronous circuits do have disadvantages as well. The clock rate (and ultimately the performance) is limited by the slowest delay path in the machine. Also, due to differences in propagation delays, trace lengths, and other factors, the clock cannot be distributed instantly to all parts of the circuit. It might arrive early in some parts and late in others. The maximum magnitude of this time difference is called the *clock skew*. This error term shortens the effective clock period, and hence reduces the circuit's overall performance.

An asynchronous circuit does not use a clock to keep itself in lock-step. Operations flow one from another, and state changes within the circuit will rarely occur simultaneously or at regular intervals. One advantage of asynchronous circuits is that clock skew errors are eliminated, and often, so is the entire clock distribution problem. Also, circuit elements can operate at their own speeds, and progress is not limited by the slowest delay path.

Asynchronous circuits are usually more complex than a synchronous equivalent. Signals may serve as strobes, and so care must often be taken to ensure that they are clean and glitch-free. Signal delay paths can be long and have many branches. This makes timing analysis difficult and less precise. Also, because signals can change at any time, this brings with it the risk of mis-sampling and metastability. Synchronizers can be used to reduce this risk, but can never eliminate it. Synchronizers also reduce performance and add circuit complexity. If timing errors do occur, they can be almost impossible to find, and debugging asynchronous logic is often more of an art form than a science.

Most bus interfaces are mostly asynchronous. This is because of difficulties associated with distributing an accurate clock, and because of the desire to avoid locking down system performance to the least common denominator.

The SBus has taken a different tack, though, and is a synchronous bus. Its small physical size and limited electrical loading make distributing a low-skew clock straightforward. Also, clock rates for the low-cost, low-power CMOS technologies that the SBus is optimized for are probably at or near their peak already. Performance advances are more likely to be achieved through increased levels of integration and parallelism, not by scaling clock speeds. As a result, fixing clock rates at a level which is aggressive today is not likely to prove a painful bottleneck tomorrow. For these reasons, there are few disadvantages to purely synchronous protocols, and the advantages they afford in simplicity, reliability, and efficiency make them well suited to the SBus' role as a chip-to-chip interconnect. Engineers who have worked with both the SBus and asynchronous interfaces (such as VME), invariably claim that SBus' synchronicity is one of its primary advantages.

Most SBus signals are sampled at the rising edge of the clock, as shown in Figure 3.2. Only the interrupt lines are asynchronous, because this simplifies sharing them among devices. The clock rate can vary from 16.67 to 25 megahertz (MHz). All synchronous signals must be sampled synchronously. Using them "raw," as either a level or strobe, is incorrect and dangerous. Their value is only guaranteed for 15 nanoseconds (ns) before the clock edge and for 0 ns after. At other times they can float, glitch, or assume any logical value.

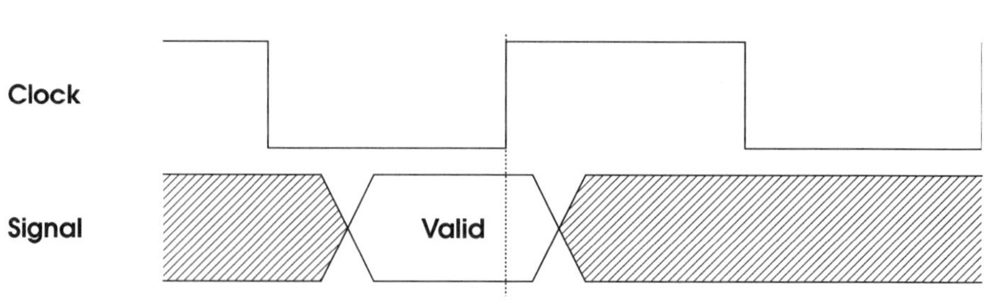

FIGURE 3.2. *Most SBus Signals are Sampled on the Clock's Rising Edge.*

3.1.2 Active Drive

ACK* and **LErr*** are control signals that are shared among all devices on the SBus. They are *low-true* signals, driven low when asserted. Between accesses they are not driven, but they cannot be allowed to float, either. If they did, they might float into an asserted state, and this could interfere with a subsequent cycle. They might also hover at or near the thresholds of the receivers, and this could increase power consumption. A more detailed discussion of this issue and its effect on other SBus signals is found in section 3.6.3.

Pull-up resistors are the traditional solution to this kind of situation, and they will also work here as long as their value is high enough. Otherwise the current levels required to overdrive them would exceed that possible for a chip-level interface. If the resistor value is high, though, the time constant it forms with the signal's capacitance will be long. The resulting rise time could be up to 2 microseconds (μs) for a line with a 10k pull-up (typical of the SBus) loaded to the maximum 160 picofarads (pF) allowed. That is tens of clock cycles, and is ultimately not much better than just letting the lines float.

Fortunately there is an easy solution that can hold these signals de-asserted between transfers, with suitably high edge-rates, but without excessive power dissipation. Called *active-drive*, it is diagrammed in Figure 3.3. The principle requires that any output that actively asserts the **ACK*** or **LErr*** signals must then actively *de-assert* them before releasing them. Then, even high-value pull-up resistors can hold them in the de-asserted state until once again driven.

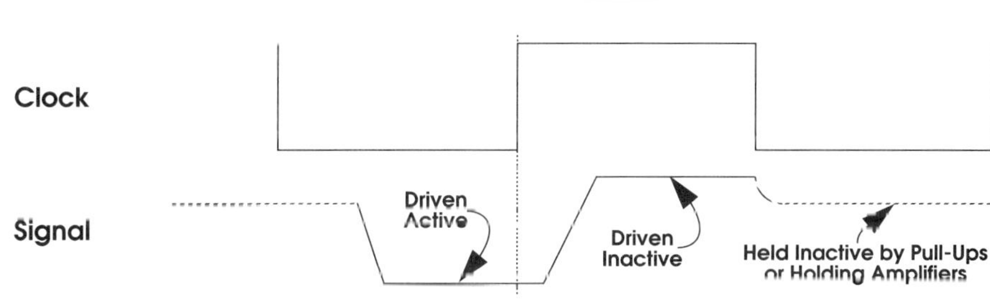

FIGURE 3.3. *Shared (Tri-stateable) Control Signals Must Be Driven Inactive.*

Note that there is no minimum time that the signal must be held inactive—it just needs to get there. This diagram shows it being de-asserted for only part of a cycle, although more commonly the process will be synchronous and the de-assertion will last for one full clock cycle. The signal should not be held inactive for *more* than one clock in any event, because otherwise this might interfere with a following cycle. It is also important to guarantee that when the signal is tri-stated it is not momentarily glitched in the process. More information on how this might occur (and on how to prevent it) is contained in section 5.1.2.

The interrupt lines must not be actively de-asserted. These lines are shared, and any attempt to forcibly de-assert them might adversely effect another device's interrupt.

3.1.3 Built-in Driver Turn-around

Many SBus signals can be driven alternately by several different devices. Ownership may change between transfers, or even during a transfer. The **Data** lines during read transfers provide a good example: At the start of a transfer the master drives the virtual address onto them, but by the end of the transfer the slave is driving the read data.

Changing ownership is not a trivial matter though. It is very important that only one driver be enabled at all times. Otherwise, a 'bus-fight' will occur. Bus fights, or driver overlaps, can cause improper logic levels, excessive power dissipation, and oscillation. This oscillation can be very dramatic if its frequency approaches the resonant frequency of the line (due to its distributed inductance and capacitance). In such cases, voltage levels exceeding 20 volts (V) at 30 MHz have been seen in a circuit with only 5-V inputs! Even in more modest circumstances, the undershoots and over-shoots that result can cause a chip to misbehave, latch-up, or fail altogether. CMOS logic is especially susceptible to these problems in general, and ASICs particularly so. As a chip-level bus empha-sizing just these technologies, then, SBus must prevent driver-overlaps from occurring under any circumstances.

Preventing bus fights is done by guaranteeing that the current driver is completely disabled before the next driver can be enabled. To do this often requires tight control over the minimum time needed to enable a driver, and the maximum time needed to disable it. These limits can clearly be contradictory, and can pose a very difficult problem both for the design and the timing analysis. This is especially true if a single reference is used, and the minimum enabling time must therefore be greater than the maximum disabling time! Multiple references could solve this problem, but this can add signals and errors. Variation of propagation delay due to process variation is another major source of error, and all errors add variables to the problem and make it more difficult to solve.

If this problem is solved in a proper worst-case fashion, then no overlap will occur even if disabling time is maximized and enabling time is minimized. What about the other extreme, though, where the disabling time is at a minimum and the enabling time is at a maximum? In this case there is a gap between the end of one driver's reign and the start of the next's. If variations are large, then the length of this gap can be, too. The lines remain undriven for all this time (pull-ups, pull-downs, or holding amplifiers will prevent them from floating, as this causes other problems). This will not 'break' anything; the bus will still work. For this reason these times are often not bounded (if they were, that would complicate the original problem even more). Still, there is a performance impact because the lines aren't useful while un-driven.

SBus' synchronous nature helps out here as it does in other aspects of the design. SBus' clock provides a reliable, repeatable reference with effectively no error (clock skew is very small, and even that can be neglected here because it is accounted for in other ways, which will be discussed later in section 3.6.4). The protocols have been designed so that no more than one output ever need drive any signal during the same clock cycle, as shown in Figure 3.4. Driver A is disabled on one clock edge, and driver B is enabled on the next.

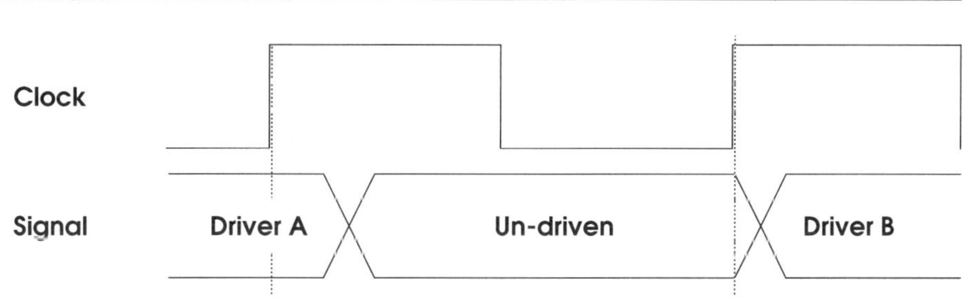

FIGURE 3.4. *No Signal is Driven by More than One Output During any One Clock.*

Given this rule, the minimum enabling or disabling times can both be 0, and are therefore easily guaranteed. Also, virtually the entire clock cycle is available to disable the driver (the specification allocates all but 5 ns for this purpose). Even at clock rates of 25 MHz this is plenty of time for the technologies used. Maximum enabling time is limited not by driver-overlap concerns, but by setup time requirements (also to be discussed later in section 3.6.4).

There is still a time when the signal can be un-driven, as also shown in Figure 3.4. It was not possible for SBus' designers to eliminate this, but it was possible to limit the ultimate impact it has on bus performance. First, there is a bound on the maximum time the line remains idle, which is a function of the clock rate. Second, the protocol was designed to minimize the number of times an ownership change is necessary, without unduly increasing the number of signals (this is one reason SBus does not multiplex the data with the physical address). Most importantly, though, the protocol has been designed so that when an ownership change is necessary, whenever possible it occurs at a time when the signals involved must sit idle anyway.

As a result of these allowances made in the SBus protocols, avoiding driver overlap problems has become a relatively simple matter. Designers need only work to reduce the maximum enable and disable times of their drivers, until the times are within acceptable limits.

3.1.4 Direct Virtual Memory Access (DVMA)

The SBus' integral support of virtual memory might seem to complicate the SBus, which would be contrary to its goals both for simplicity and as a chip-level interface. Actually the opposite is true, however; virtual memory support *simplifies* the SBus and reduces the amount of logic necessary to provide a high-performance device capable of *Direct Memory Access* (DMA).

Direct Virtual Memory Access (DVMA) is a simplification for several reasons. First, the number of address bits that must be handled is reduced because the device need only address the size of the space it needs, not the entire memory space. More importantly, though, the address space can be made to appear contiguous, whether or not it is actually fragmented or widely dispersed. This allows sophisticated DMA operations such as scatter-gather and chaining to be accomplished without the need for additional hardware or frequent, complex initialization. Efficiency is improved because the hardware and software tools necessary to manage

memory are provided by the host from the outset, and do not need
to be duplicated or re-invented for each new application.

DVMA also provides substantial benefits in terms of autoconfigurability, because it allows devices to be soft-mapped. Autoconfiguration was also a major goal of the SBus, and something
that simplifies the hardware (because no DIP-switches or other
mechanisms are needed for configuration) and the software
(because no addresses need to be hard-coded).

3.2 Components

Every SBus transaction has three participants. An SBus *master*
interface initiates the transfer request. A *slave* interface performs
the operation. The SBus *controller* arbitrates transfer requests,
translates virtual to physical addresses, and oversees the whole
process.

Physically, any one device can contain one, two, or all three of
these components. For example, the host motherboard in a SPARC-
station class machine contains the controller, a master interface,
and a slave interface. Most add-on cards which contain master
interfaces also contain slave interfaces. Logically, though, these
elements are separate, independent entities.

3.2.1 Master

SBus masters initiate SBus transfers. They usually contain processors or advanced state machines, which provide them with the
intelligence required to satisfy their own data transfer needs. The
SBus supports multiple masters because this reduces the burden
that would otherwise fall on the host CPU (which usually contains
its own master interface).

In addition to initiating the transfers, masters are also
responsible for supporting slaves of varying widths (through
dynamic bus sizing) and for error recovery and reporting (should a
transfer fail).

3.2.2 Controller

The SBus controller is responsible for a wide variety of functions.
It is the source of the SBus' clock. It arbitrates the bus request signals from each of the masters and provides a bus grant to whichever master it decides has next access to the bus. After the master
begins the cycle, the controller is responsible for translating the

virtual address into a physical address, which it then drives onto the bus. It also does some of the address decoding, generating a select signal to whichever slave is the target of the current operation. The controller also generates a strobe that indicates both the address and select signal are valid, and that the transfer phase of the transaction has begun.

After the transfer phase has begun the SBus controller becomes primarily an observer. It examines and counts acknowledgments to determine when a transfer has finished, then de-asserts the address strobe and select signals, then grants the bus to the next requesting master (if any). The controller intercedes in the current transfer only if no response is evident within the preset time limit of 256 clocks. In that case, the controller provides the response instead, with an error acknowledgment.

3.2.3 Slave

SBus slaves are the simplest of the three participants. This is both useful and intentional, because slave interfaces are also the most commonplace. An SBus slave can only respond to requests generated by an SBus master. It cannot initiate requests, except indirectly by generating an interrupt.

SBus slaves can vary in width. Valid widths are 1, 2, 4, or 8 bytes. The width chosen for any given device is a tradeoff between the design's complexity and bandwidth it requires. Bus sizing allows data to be transferred to slaves that are physically narrower than the data's width.

3.3 Configurations

There are a number of possible architectures for SBus-based systems. Three of the most common are host-based, symmetric, and multichannel. There are other possibilities, and the actual architecture of any one system might prove a mix of elements from several of these. The three previously mentioned architectures will encompass the vast majority of systems, however, and so are discussed here.

3.3.1 Host-Based

The architecture of a host-based SBus system is shown in Figure 3.5. For simplicity's sake, the SBus controller is not shown, although the Memory Management Unit (MMU) that it uses for virtual address translation is shown. (This MMU can be dedicated to the SBus, or it can be shared by the cache/memory subsystem.)

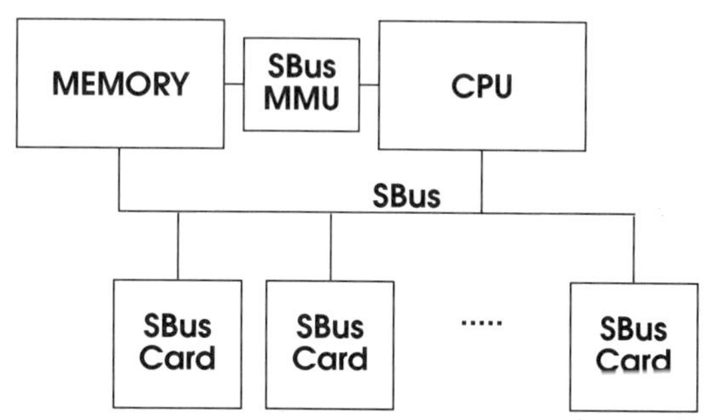

FIGURE 3.5. *A Host-Based SBus Architecture.*

The key distinguishing feature of a host-based configuration is that one or more SBus masters in the host system have "back-door" access to the SBus' MMU. This allows them to perform address translations in parallel with other SBus activity, so that they do not need to perform the address translation phase of an SBus transfer. Instead, the SBus controller immediately issues a physical address when granting the SBus to the host, following with **AS***, and orchestrating the remainder of the transfer as before.

Such architectures can reduce the latency and increase the bandwidth of the SBus, not just for the host, but for all devices. This is because the bus cycles it would have used for address translation are now available for other purposes. The host typically initiates a large percentage of SBus transfers, so this savings can be substantial.

3.3.2 Symmetric

A symmetric architecture differs from a host-based configuration in that no master has any special path to the SBus. All transfers are treated identically, and each master must use the bus to perform the virtual address translation. A diagram of such an architecture is shown in Figure 3.6. Once again, the SBus controller is not shown for simplicity's sake.

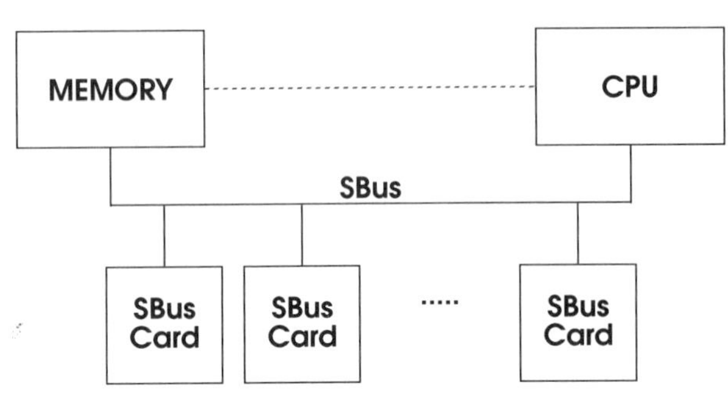

FIGURE 3.6. *A Symmetric SBus Architecture.*

This kind of architecture is very simple, logically and physically. It will often include some kind of private path (shown here by the dashed line) between the processor(s) and system memory, because the SBus is not well suited for this purpose.

3.3.3 Multi-Channel

One of the most common criticisms of the SBus revolves around the small number of slots available. Most machines have only two to three slots, which seems woefully inadequate to those coming from a PC environment where between five and eight slots are commonplace. Electrical limits do prevent much beyond about four slots, but SBus's small size and electrical simplicity provide a ready solution. Multiple SBus interfaces can be attached to any one host, as shown in Figure 3.7. For simplicity's sake the SBus controller is not shown, but each of the four SBus channels shown would have its own controller function.

There are significant advantages to this kind of approach. The number of slots on any given machine can be increased as necessary without increasing the electrical loading, capacitance, and routing lengths. The buses can also operate independent of each other, multiplying the I/O bandwidth available to the host.

One limitation of such a multi-channel approach is worth mentioning. On any one SBus, any master can communicate with any slave. Neither the CPU nor system memory need to take part in every transaction. This would remain true with multi-channel

SBus architectures as well, except that such peer-to-peer communication would not be guaranteed between a master on one bus and a slave on a different bus. The system's designers may provide this feature, with some form of cross-point switch or other arrangement, but they are not required to do so.

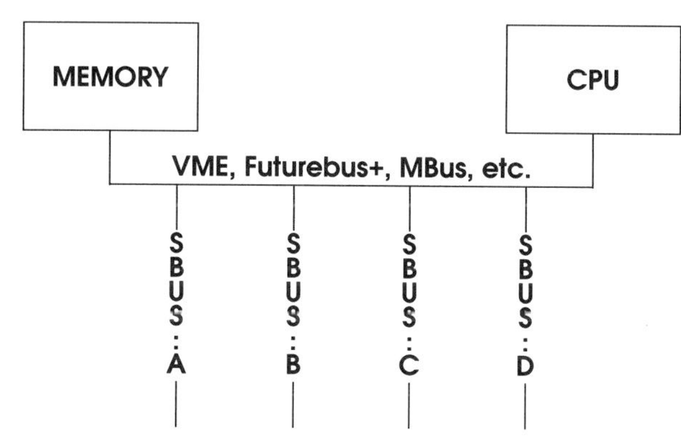

FIGURE 3.7. *A Multi-Channel (or Parallel) SBus Architecture.*

Except for the multiple SBus channels, this system's architecture might otherwise be either host-based or symmetric.

3.4 Protocol

3.4.1 Signal Summary

The following section contains a summary of all the SBus' signals. Again, all but the interrupt signals are synchronous, and their value depends not just on their electrical level, but on their timing relative to the clock signal.

Clock

The **Clock (Clk)** signal is generated by the controller. It is a constant frequency that must lie between 16.67 and 25.0 MHz. The rising edge of the clock provides the SBus' timing reference and should be used to sample synchronous signals. The falling edge of the clock should not be used as a (mid-cycle) time reference because there is no duty cycle restriction. The clock must remain high or low for at least 17 ns, however.

This signal's rise or fall time is limited to at least 1 nanosecond and at most 3 ns. In part this requirement is necessary to minimize the clock skew (or distribution error), which must be no greater than 2.5 ns.

Reset*

The **Reset*** signal is generated by the controller, and is used to reset all SBus devices at power-on or at any other time necessary. It must be asserted for at least 512 clock cycles after power is stable. It will be de-asserted synchronously, but there is no guarantee that it will be asserted synchronously. In fact, there is no guarantee that it will be obviously asserted at all (see the related discussion starting on page 151).

There is no minimum time guaranteed from **Reset*** de-assertion to the first SBus access, and designers should not make any assumptions about this. If your card contains logic that takes a finite time to initialize after reset, "hard-code" in enough intelligence to issue rerun acknowledgments until fully initialized. See page 152 for a detailed discussion of this issue.

BusRequest* and BusGrant*

The **BusRequest*** (**BR***) and **BusGrant*** (**BG***) lines are used together to perform arbitration for control of the SBus. These signals are radial; there is one unique pair of these for each and every SBus master in the system. The SBus master asserts **BR*** when it wants access to the bus, and the SBus controller asserts the corresponding **BG*** when the bus is available for that master's use.

PhysicalAddress(27:0)

The SBus controller drives the **PhysicalAddress(27:0)** lines (**PA(27:0)**) with the results of the virtual address translation it performs for the SBus Master. Not all SBus controllers drive all 28 of these lines. Some drive only the 25 signals which comprise **PA(24:0)**. In this case, the unused **PA(27:25)** can be pulled up, grounded, or in some early designs even allowed to float.

In most cases **PA(27:0)** need not be latched on the slave; they are held stable by the controller for the duration of the transfer. During extended mode transfers, however, the slave must latch the address because it is valid only during the clock in which **AS*** is first asserted. After that, these signals become **D(59:32)**.

SEL* and AS*

In addition to translating the virtual address, the controller does all address decoding necessary to determine exactly which slave will be participating in the upcoming transfer. Each slave has its own, unique **Select*** (**Sel***) signal. After the controller has determined which slave is being addressed, it asserts that slave's **Sel***.

When the physical address is valid and the appropriate **Sel*** signal is asserted, then the controller asserts the **Address Strobe*** (**AS***) signal. Slaves must use this signal to qualify **SEL***. These two signals should be considered inseparable. Only under the rarest of circumstances should one be used without the other, and they should always be used synchronously. A more detailed discussion begins on page 148.

Read

The **Read** (**Rd**) signal is driven by the SBus master. When asserted during a transfer, this signal indicates that a read operation is requested. This transfers data from the slave to the master. A write operation, which transfers data from the master to the slave, is indicated if this signal is de-asserted during a transfer.

In most cases **Rd** need not be latched on the slave; it is held stable by the master for the duration of the transfer. During extended transfers, however, this signal is always de-asserted during the clock in which **AS*** is first asserted (and then held there with pull-downs until over-driven). This seemingly indicates a write transfer, but is actually a mechanism to help ensure backward compatibility (see section 3.4.6 for more detail). The extended transfer's actual direction is encoded in the Extended Transfer Information Word, which is multiplexed onto **D(31:0)**.

Data(31:0)

The **D(31:0)** signals (**D(31:0)**) are primarily a data path, but they are also used to transfer virtual addresses to the controller for translation, and the Extended Transfer Information Word on extended (64-bit) operations. During any given transfer these signals can be driven by the master, the slave, or both.

Due to byte placement requirements (discussed starting on page 64), SBus masters must be connected to all 32 data signals. Slaves can be either 8, 16, or 32 bits in width, and need connect only to those signals they need.

Size(2:0)

The master encodes the size of the transfer it is to perform on the three **Size** (**Siz**) lines. The encoding used is shown in Figure 3.8. These signals are positive-true, and so the values listed in the table will exactly correspond with the signal levels that would be found if these signals were probed in a real machine.

Siz(2:0)	Function
000	Word (4 bytes) transfer
001	Byte transfer
010	Half-word (2 bytes) transfer
011	Extended Transfer*
100	Four-word (16 bytes) burst
101	Eight-word (32 bytes) burst
110	Sixteen-word (64 bytes) burst
111	Two-word (8 bytes) burst

*The actual transfer size is encoded within the Extended Transfer Information Word

FIGURE 3.8. *Size(2:0) Signal Codes.*

A slave need not and should not distinguish between non-burst sizes that are greater than or equal to its port width. Bus sizing will be indicated if the transfer size is greater than the slave's width, and this is transparent to the slave. In other words, a byte-wide slave can respond equally well to byte, half-word, or word requests. A half-word wide slave can respond equally well to half-word, or word requests. For a more detailed discussion, see the discussion on byte ordering and placement which begins on page 64.

In most cases **Sel(2:0)*** need not be latched on the slave; these signals are held stable by the master for the duration of the transfer. During extended mode transfers, however, these signals broadcast the extended transfer code only just before and during the clock in which **AS*** is first asserted. After that, these signals become **D(62:60)**. Note that the extended transfer size code indicates only the operation's intended width. The transfer's actual size (number of long words) is encoded in the Extended Transfer Information Word, which is multiplexed onto **D(31:0)**.

Unlike all other SBus signals, the interrupts are asynchronous, and can change at any time in the clock cycle. This is because most of the potential interrupt sources (such as an I/O device) are already asynchronous. Allowing the interrupts to remain asynchronous until they reach the processor saves logic because only one synchronizer is needed at the destination, where one *per source* would be necessary otherwise. Reliability also improves because synchronizing at the destination is more effective than synchronizing at the source: clock skew is eliminated and the slow rise times, electrical noise, and glitches inherent in an open-collector environment are filtered out.

A consequence of the open-collector nature of these interrupt signals is their relatively slow rise time. It is therefore recommended that Schmidt trigger inputs (or other inputs with built-in hysteresis) be used to receive these signals.

DataParity

The **Data** lines can optionally be parity checked using the **DataParity** (abbreviated **DtaPar**) signal. This signal is driven by the same master or slave that is driving the **Data** lines, and must be driven or tri-stated at the same times that the **Data** lines are. This signal checks **D<31:0>** during all parts of standard 32-bit mode transfers, and during the translation phase of extended 64-bit transfers. During the remainder of all extended 64-bit transfers **DtaPar** checks **D<63:0>**.

The parity signal is driven so that the total number of signals in the checked group (including **DtaPar**) is an odd number. This is called *odd parity*. Another way of stating this is that the exclusive-or (XOR) of all checked bits must equal a logical '1'.

The card can *generate* parity at all times, whether or not it is supported by any other part of the system. It must only *report* parity errors, though, if the data's source also generates parity. Any card capable of generating and checking parity must contain an FCode attribute named **parity-generated** in its ID PROM. The device drivers involved in the transfer must look for this attribute and enable parity error reporting only if it is found.

Parity support is optional, and because of the complexity involved it should only be considered if data integrity is of utmost importance. A common misconception is that parity increases reliability. This is not true, and in most cases just the opposite occurs. The added complexity and logic required to support parity adds failure modes and can actually increase the failure rate (though usually by only a small, perhaps insignificant factor). What parity can do is help find any failure that does occur *earlier*, hopefully before it has had a chance to corrupt data.

LateError*

The **LateError*** (**LErr***) signal is an alternate way to report errors, without generating an error acknowledgment. This signal's name implies its primary purpose: it is used to report errors *after* an acknowledgment has already been issued. This is useful when parity or error-correction circuitry is involved in a transfer. This type of logic might take a relatively long period of time to produce a result. Rather than delay the acknowledgment (and hence lengthen the SBus cycle) until all results are tallied, it is possible for the slave to generate an acknowledgment assuming that no errors will occur. If this assumption is later proven false, though, the slave can assert **LateError*** to indicate this.

Any SBus slave can drive **LateError***, and all masters must sample it. The timing of this signal is critical. It is only sampled exactly two clock edges later than the acknowledgment was sampled (see the timing diagrams in the following section).

3.4.2 Basic Transfers

The previous section introduced the SBus' signals and some preliminary information on their purpose and use. This section provides even more detail, especially on how the signals interact with each other.

All SBus transfers can be broken down into three major stages. The first of these is the *Arbitration Phase*, during which SBus masters request and can be granted access to the bus. The next is the *Translation Phase*, during which a virtual address is translated into a physical address. The final stage is the *Transfer Phase*, where the data is actually transferred between master and slave. Each of these phases is discussed in the remainder of this section.

Three closely related topics are also contained in the following discussions. *Flow-Control* allows the slave to throttle its throughput rate. *Dynamic Bus-Sizing* is a mechanism that allows slaves of varying port-widths in a way that is largely transparent to software drivers. *Byte Ordering and Placement* defines the relationship between data lines, access sizes, and addresses. It also describes the effect that bus-sizing has on this relationship.

A basic SBus transfer is presented as a timing diagram in Figure 3.10. This diagram, and the related ones that follow, use the same drawing conventions used in revision B.0 of the SBus Specification (and illustrated in Figure P-1. of that document). Extended Mode (64-bit) transfers behave differently. These are discussed in more detail in section 3.4.6.

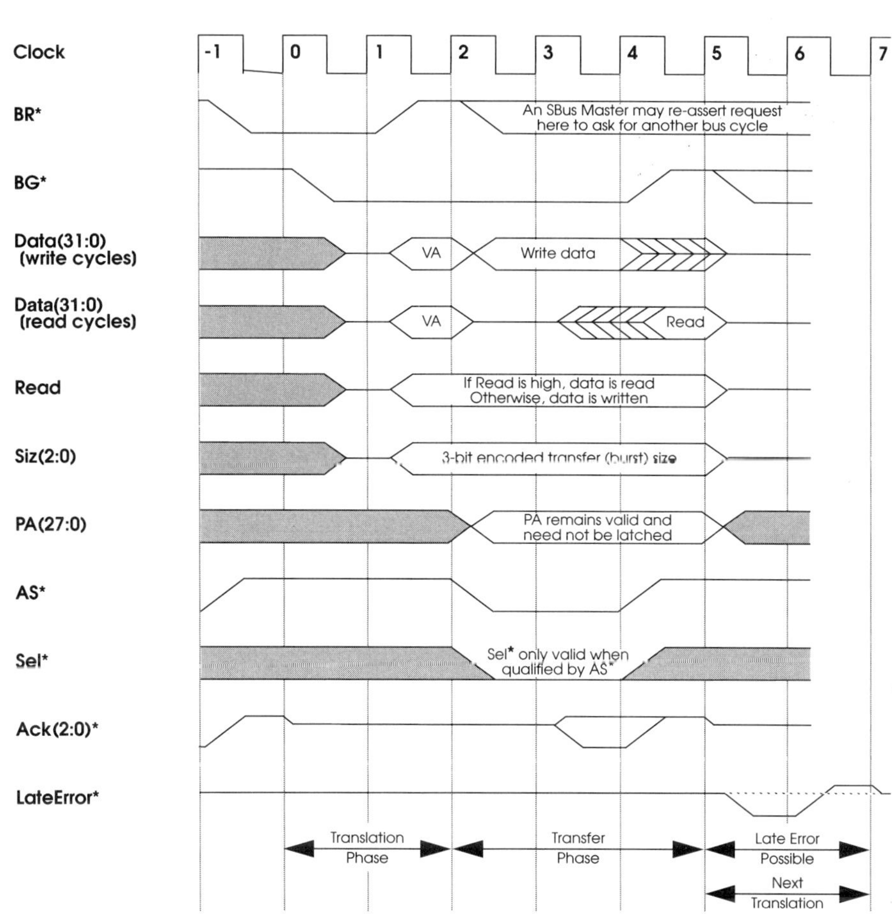

FIGURE 3.10. *Basic SBus Transfer.*

Arbitration Phase

The SBus supports multiple masters, each of which can initiate transfers. Occasionally, more than one of these masters will want access to the bus at the same time. This is impossible, of course, and so the SBus controller must decide which master will perform the next transfer. This occurs during the arbitration phase. Each master has its own unique signals dedicated to this purpose, and it is possible to perform arbitration for the next transfer while the current transfer is still in progress. This is important, because it means that there is no performance lost to this overhead function.

An SBus master indicates that it wants to perform a transfer by asserting its **BusRequest*** signal. There is one of these signals for each potential master, and so the SBus controller is immediately aware of which master has made the request and now wants to transfer data.

If this is the only request that the controller currently has pending, then no arbitration is necessary. If, however, multiple **BR*** signals are active at any one time then the controller must choose which master will be able to perform its transfer next. The method it chooses to do this can be based on priority, order-of-request, or any other appropriate mechanism. The specification only requires the arbitration method be "fair."

In any event, the master is notified that it may proceed with its transfer when the controller asserts its **BusGrant*** signal. Like the **BR*** signals, there is one **BG*** for each potential master.

Normally, the master must release **BR*** for at least one cycle when it receives **BG***. It can then immediately re-assert **BR*** if it wishes to do another transfer after it finishes the current one. There is a special significance if the SBus master does not de-assert **BR*** for at least one cycle, but holds it asserted even after **BG*** is received. This is the mechanism used to indicate that an atomic cycle is in progress. Atomic cycles are covered in detail in section 3.4.4 below.

Translation Phase

As previously mentioned, SBus transfers are based on virtual addresses. These virtual addresses must be translated into physical addresses and then partially decoded before the proper slave can use them. This is the controller's responsibility, and it occurs during the translation phase which leads most SBus transfers.

The translation phase begins after the master samples **BG*** and finds it asserted. The master must then immediately drive a virtual address onto the **Data** lines, the transfer size it wishes to perform onto the **Siz** lines, and the transfer direction onto the **Rd** line. All these lines must be driven quickly enough to meet the setup time requirements for the next clock edge (clock #2 in the timing diagram). The virtual address is held for exactly one clock.

After that, the master must either replace it with the data to be transferred (if a write) or tri-state the lines (if a read). The **Siz** and **Rd** lines must be held stable throughout the transfer.

The SBus controller then translates the virtual address into a physical address. The method it uses to do this can vary from one machine to the next; it is not specified. The only requirement is that the MMU must provide the capability of mapping pages which are no greater that 64 Kbytes in size. It may also be capable of mapping pages larger than this, as long as page mapping within this limit can be accommodated at the same time. The reason for this restriction is to allow the MMU to be used as a mechanism which can restrict access to certain portions of a device's address space. Normally, this is used to write-protect areas. The maximum page size that all systems are able to support is a crucial piece of information when dividing a device's address space. The designer is guaranteed that the MMU can provide separate mappings (and hence protections) for any given 64 Kbyte block. On some machines the granularity is finer than this, but the card designer must plan for the worst case, not the best case.

Once the translation is complete, the controller drives the physical address onto the **PA** lines (only one clock cycle is required for the translation in this diagram, but this is not a requirement and additional cycles may be used by the controller if necessary). It also partially decodes the address and asserts one of the **Sel*** signals to the SBus slave which is the target of this transfer. Like **BR*** and **BG***, the **Sel*** signals are unique to each slave. Therefore, the slave knows immediately and without additional decoding that it is expected to participate in the current transfer.

Once the physical address and **Sel*** signals have been driven, the controller then asserts **AS*** to indicate the validity of this information. **Sel*** must only be sampled synchronously, and both it and the physical address are only valid when **AS*** is (also synchronously) asserted.

Translation phases may not occur in some SBus transfers on host-based SBus architectures (discussed above). This is because the CPU master may have special "back-door" privileges with the controller that allow it to perform translations in parallel to other SBus activity. In this case, only the transfer phase will be visible on the bus.

Transfer Phase

The transfer phase is that part of an SBus transaction where data is actually moved to or from the slave. The slave is the lead-performer in this part of the operation, and for that reason this phase has also been called the *slave cycle*. The term "cycle" can be misleading, however, because it is often used in reference to individual clock periods, of which there may be several in the transfer phase.

If the current operation is a write, the master provides and holds the data on the **Data** lines until the slave either is ready, signals an error, or asks the master to retry the operation. The slave does this via an encoding on the three **Ack*** lines, discussed in the preceding signal summary. If no slave has responded within a set time-out period then the controller must assert an error acknowledgment, thus ensuring a deterministic outcome. The master must guarantee that the data remains valid until the clock edge on which the acknowledgment is asserted (clock #4 in Figure 3.10.). The herring-bone pattern indicates that the master may continue to drive data for one clock past that in which the acknowledgment is sampled. In some applications this may simplify the master's driver-enable logic.

If the current operation is a read, the slave must place valid data onto the **Data** lines after the acknowledgment is asserted. This data must be held until the master samples it on the clock edge immediately after that in which the acknowledgment was asserted (in this case, the rising edge of clock #5).

The slave may indicate an error on the transfer either by asserting an error acknowledgment, or by driving **LateError***. If the latter is used it will be sampled exactly two clock edges after that in which the acknowledgment was asserted (in this case, the rising edge of clock #6).

After the last acknowledgment is asserted in any transfer, the controller de-asserts **AS*** and the current **BG*** (this is not guaranteed to occur immediately and may take several clocks). The controller may not assert a new **BG*** at the same clock edge on which the previous one is de-asserted; all **BG*** signals must remain de-asserted for at least one clock. This is necessary to prevent driver contention on the **Data** lines.

Flow Control

SBus slaves can control the data transfer rate by simply controlling the rate at which it issues data acknowledgments. This is often referred to as flow control. The slave may issue its acknowledgment as soon as it is possible, minimizing the transfer's time. Or it may insert wait-states if it requires time to retrieve or store the data, perform a calculation, or secure access to a resource. On burst operations, the slave may issue data acknowledgments on successive clocks, allowing data to be transferred at a rate of one word per clock. Or it may space out the acknowledgments to suit its needs or capabilities.

SBus masters have no equivalent mechanism and must be able to provide data at whatever rate the slave is capable of accepting it. This means that the master must be able to provide data at the maximum word-per-clock rate.

Dynamic Bus-Sizing

By now there have been several references to dynamic bus-sizing, which allows a master to more easily communicate with slaves of varying widths. This is a purely hardware-based mechanism, and is therefore transparent to both firmware and software.

Slaves of varying widths are allowed because it can greatly reduce their complexity and cost. It reduces the number of signals that must be driven or received, which often means fewer or smaller parts can be used. Simplifying SBus slaves is an important goal, even if it comes at the expense of the master interface. This is because every board must contain a slave interface (if only for the ID PROM). Relatively few devices, however, require DVMA (master) interfaces. For these reasons, the master interface must assume the primary responsibility for bus-sizing.

Neither the master nor the slave is required to support bus sizing, however. The master is free to abandon a transfer that cannot be completed at the size originally requested. The slave may also choose not to cooperate. It can respond to any transfer request larger than its port width with an error acknowledgment, although there is little or no reason to do so (see the discussion 'Avoid Restricting **Siz** Codes,' starting on page 262).

Bus sizing occurs as the result of an implicit negotiation that occurs between the master and the slave during a transfer. The master indicates the size of transfer it wishes to perform via the code it places on the **Siz(2:0)** lines. The slave responds with its port width, which it encodes on the **Ack(2:0)** lines. If the two don't agree then only a part of the data requested is transferred. The master must perform additional *follow-on* cycles to transfer the remainder. these follow-on cycles are separate, distinct, non-atomic transfers from the slave's perspective.

For example, suppose a master initiates a 32-bit transfer to an 8-bit slave. The slave will provide (or store) the first byte of data and an acknowledgment which indicates its width. A master which supports bus-sizing must then perform three additional transfers to move the remaining 3 bytes. The master might change the **Siz** code to indicate that the follow-on transfers are byte-operations, but it is not required to do so. Instead, for simplicity's sake it may continue to provide the 32-bit request. The address must be adjusted as appropriate, however, and the master must also adjust the data placement. This is discussed in more detail in the following section.

Dynamic bus-sizing only provides support for slaves of varying widths. It cannot be used to adjust the lengths of burst transfers. If the slave is not capable of handling the burst size requested then its only recourse is to issue error acknowledgment.

Note that a slave's port width may vary with address. For example, in some regions of its address space it might behave like an 8-bit device, and in other regions it might behave like a 16-bit device. Any such change must not occur within any one word. In other words, all byte or half-word locations within a word must respond with the same type of acknowledgment. This is to assure the SBus master that if bus-sizing occurs, then all possible follow-on cycles will provide a consistent acknowledgment. Otherwise, the resulting confusion would complicate the master.

Byte Ordering and Placement

On the SBus, a byte's significance decreases as its address increases. For example, the byte at address 0 is more significant than the byte at address 1, and so on. This is shown another way in Figure 3.11. Here, a sample 32 bit word is broken up into each of its 4 constituent bytes. Also shown are the two least-significant address bits associated both with the word and with each of the bytes. The data bits which represent the data's normal or *natural* location are identified, too. These data bits form the *byte lane* associated with that address.

Data is not always transferred in its natural location, however. Slaves which are narrower than the SBus' 32-bit data width cannot use data lines which they are not connected to. A byte-wide slave must transfer all data on the single byte lane to which it is attached, for example. A half-word-wide slave must transfer all data using only the two byte lanes to which it has access. The actual byte lanes used on any given transfer is therefore a function both of the address and the slave's acknowledgment (which contains information about its width).

This function is designed with a very simple premise. Data will be transferred wherever it is most convenient for the slave interface. The master interface has the entire responsibility for placing or retrieving the data on the appropriate byte lanes. The reasons for this are the same as those that explain why slaves of varying widths are supported in the first place; slaves are more commonplace than masters, and hence more cost-sensitive. They must be kept as simple as possible.

Addr<1:0>	Hex Value	Data Bits
word addr: 00	FE00A543	D<31:00>
byte addr: 00	FE	D<31:24>
01	00	D<23:16>
10	A5	D<15:08>
11	43	D<07:00>

FIGURE 3.11. *As the Address Increases, the Byte's Significance Decreases.*

To complicate the master's responsibility, it does not know (and usually should not make assumptions about) what kind of data acknowledgment it will receive on the transfer. On writes, therefore, it must place copies of the data in all possible locations that it might be transferred. On reads, it must wait until it sees the acknowledgment before it picks the data from the proper data lines. This process is referred to as *byte steering*.

The function which relates address, transfer size, acknowledgment, and the byte lanes used for the transfer is graphed in Figure 3.12. There is an implicit assumption here that a slave will always respond with an acknowledgment which is equal to its port width, *even if the transfer requested is smaller.* Otherwise, the slave may be forced to do some byte-steering, too, which is counterproductive.

Note that all transfer sizes are not allowed at all addresses. This is because of the requirement that all transfers be aligned to the address. There is only one apparent exception to this rule. If bus-sizing occurs then the master may use the same **SIZ** encoding for both the initial cycle and all follow-on cycles, even though the address used for the follow-on transfers will change and may therefore seem to become un-aligned. Rather than consider these unaligned transfers, though, it is better and easier to view the situation another way. A byte-wide slave need not distinguish between 8-bit, 16 bit, or 32-bit **SIZ** codes; it may treat them all as byte transfer requests. Likewise, a half-word-wide slave can treat both 16-bit and 32-bit **SIZ** codes as if they were half-word requests. This is a good way to design slaves because it simplifies logic and maximizes the chances of finding a way to transfer data despite port width differences. Also, the slave need not know or care that bus sizing may occur, and addresses will always be aligned.

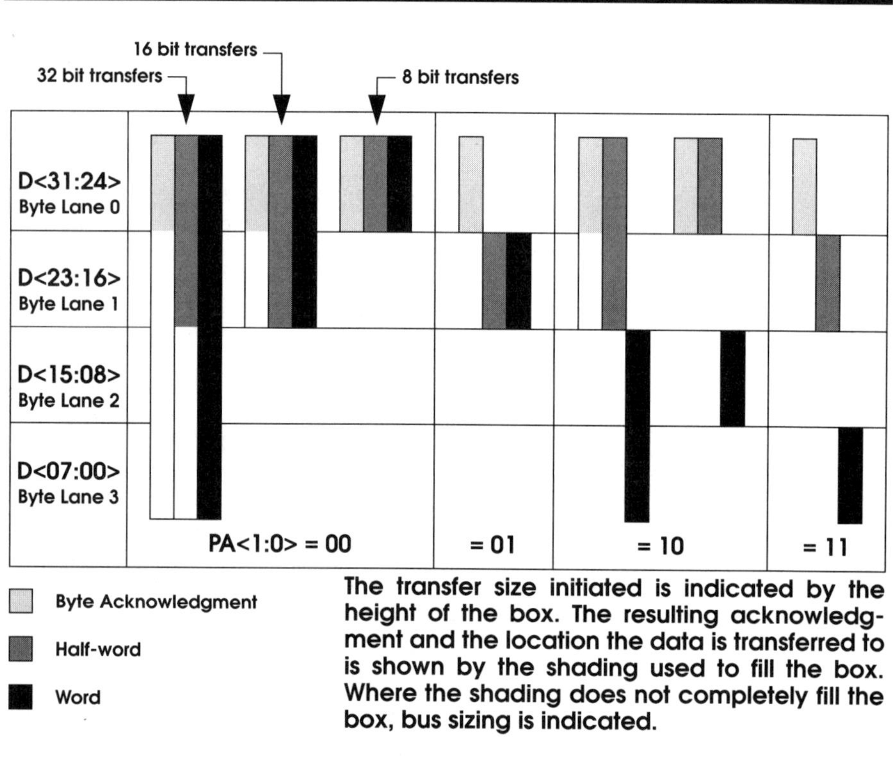

FIGURE 3.12. *Data Transfer Location is a Function of SIZ, PA, and ACK.*

For practice in using this graph, examine the case of a byte transfer where **PA<1:0>**=11. This data may be transferred to any of three possible locations. This would be **Data<31:24>** for byte-wide slaves (or wider slaves that respond with a byte acknowledgment),

Data<23:16> for half-word slaves (or acknowledgments), or **Data<7:0>** for word-wide slaves (or acknowledgments). The master must steer data to or from one of these three byte lanes as appropriate.

This particular example points out an error in the SBus Specification. The specification summarizes the master's responsibilities by requiring that the master steer data to or from two possible locations. Data must always be steered to its natural location. In addition, byte or half-word data must also be steered to or from byte or half-word lane 0, respectively. Of course, these two locations may actually be one and the same, depending on the address. This is a completely accurate summary except in the case of a half-word acknowledgment to a byte transfer where **PA<1:0>**=11. In this case the master will steer data through byte lane 0 and byte lane 3 (the natural location). But the slave will expect to transfer data on byte lane 1! Future revisions of the specification will eliminate this discrepancy, and require that in this case the master take all three byte lane possibilities into account.

So far, it must seem that byte-steering vastly complicates the design of an SBus master. This is not necessarily true, however. The specification allows the master to copy data onto all byte lanes if desired, not just those required. This eliminates the need to make an up-front decision during write transfers. Unfortunately data on read transfers must still be multiplexed only from the proper location, but there is more time available in read transfers to make the necessary choice.

The specification also allows a master interface to limit the types of acknowledgments which it will accept on any transfer; it is not forced to do bus-sizing. For example, a master may choose to accept only word acknowledgments. If it does this, then data is always transferred in its natural location, and no byte-steering is necessary at the slave or at the master! This choice will obviously limit the master's ability to communicate with slaves of varying widths, but it may be a good trade-off for masters which are designed to communicate only with word-wide slaves. There is precedent for this, too; for example, LSI Logic's L64853(A) SBus interface ASIC is optimized to perform DVMA operations only to the host system's memory. While there is no guarantee that the system memory interface is at least 32 bits wide, it is a safe assumption.

So far this discussion has not considered 64-bit transfers. This is because bus sizing (and therefore byte steering) is not defined for extended transfers. The only recourse of a slave which cannot accommodate a 64-bit transfer request is to generate an error acknowledgment. The master must then back off and transfer the data using multiple 32-bit (or smaller) transfers as described here.

3.4.3 Burst Transfers

The basic transfers discussed above move only one word (or less) of data during each transfer. SBus burst transfers are almost identical, except that multiple words are transferred. Burst transfers are more efficient than non-bursts because the overhead burden of each transfer is shared by multiple words instead of single words, half-words, or bytes.

For example, consider the case of a bus which operates at 25 MHz and can transfer 1 word (4 bytes) of data on every clock. Were there no overhead this bus could transfer data at a rate = 25 MHz x 4 bytes = 100 Mbytes/second (remember that 1 Hz = 1 cycle per second, or cps). Assume now that overhead is required, such as the four clocks that the SBus generally needs (these are detailed in Section 3.5 below). These four clocks must be added to whatever number of clocks are required for the actual data transfer.

If a single word is being transferred then the entire 4 byte operation takes 5 clocks. This data rate = (25 MHz/5 clocks) x 4 bytes = 20 Mbytes/s. This translates into only a 20% efficiency!

Consider now a burst operation which transfers 16 words (64 bytes) at a time. This operation requires one clock per word and four clocks overhead, or 20 clocks total. In this case the data rate is = (25 MHz/20 clocks) x 64 bytes = 80 Mbytes/s. This is much improved, and represents a respectable 80% efficiency. Theoretically efficiency will approach 100% with increasing burst size, but at some point the trade-offs required make large bursts undesirable. These trade-offs are discussed in Section 3.5 below.

Protocol

The sequence of events in a burst transfer is almost identical to that of non-burst operations, except that multiple words are moved during the transfer. Arbitration is performed the same way. Virtual addresses are translated the same way. The slave acknowledges each word transferred.

One other difference is that burst transfers do not allow dynamic bus-sizing or slaves of varying widths. All burst operations are word oriented, and the addresses used must be word aligned. After all, the purpose of burst transfers is to maximize the efficiency of the transfer as much as possible. As a result, there are only three acceptable acknowledgments during a burst transfer: word (double-word for 64-bit wide transfers), error, or rerun. The use of any other type of acknowledgment will produce undefined (and probably undesirable) results.

An error acknowledgment should be issued if the slave cannot perform the transfer requested, or if it encounters an error. Whenever possible the error acknowledgment should be the first acknowledgment given. However, it is conceivable that a slave might detect a parity error or some other error after the burst transfer is well underway. For this reason the slave may issue the error acknowledgment at any time during the transfer phase, even if some words have already been transferred.

A rerun acknowledgment is another possibility. This must only be issued on the first word of a transfer. (See the discussion in Section 5.3.3 on page 167.)

The other, most common, acceptable acknowledgment on burst transfers is either a word (for 32-bit wide bursts) or a longword (for 64-bit wide bursts) acknowledgment. The SBus slave must provide one such acknowledgment for each word transferred. In this way it controls the rate at which it accepts data, as discussed above under Flow Control. The slave is not required to de-assert the **Ack*** signals after each acknowledgment. This would be inefficient. Instead, the slave may drive and hold the acknowledgment code for as long as it is able to accept or supply data at the one-clock-per-word rate. It may de-assert the **Ack*** signals between transfers if it cannot keep pace. For every clock that the **Ack*** signals are de-asserted, exactly one wait-state will be added to the access.

Address Wrapping

Burst operations transfer data in blocks. An SBus master may start such an operation at any word aligned address. The slave is responsible for adjusting this address for each word transferred, because there is not time for the master or controller to update the address and pass it synchronously to the slave. The slave might also be required to *wrap* the address in a pre-defined way. This means that the address may not simply increase monotonically, but may actually jump backward once before it increases again to the end of the cycle!

One possible scenario is shown in Figure 3.13. Here the burst transfer starts at address ??24 (the most significant address bits are not important here). The transfer then logically proceeds to address ??28 (these are word addresses, remember), then ??2C. At this point something unusual happens; the address hops back to ??20, a point *before* the first word transfer.

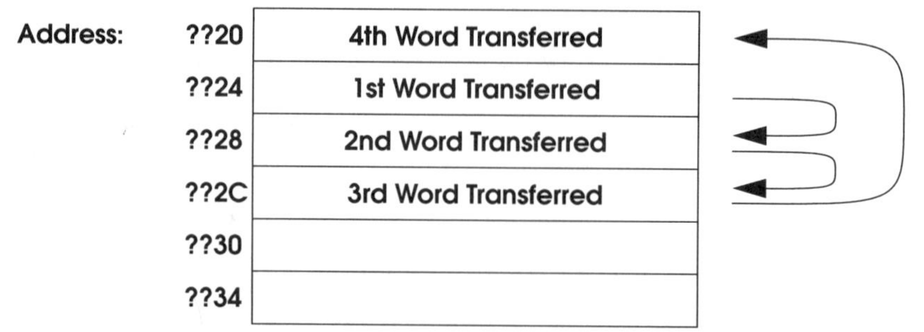

FIGURE 3.13. *Sample Address Wrapping Sequence for a 4-word Burst Transfer.*

Though this may seem odd at first, it can be a quite useful feature. This is called address wrapping, and is often used in cache or block oriented environments. Its purpose is to allow data blocks to be easily transferred without worrying about block boundaries. It also allows the word of data actually needed first to be transferred first, so that processing can proceed while the rest of the data is transferred in the background.

For example, assume that an SBus interface capable of burst transfers includes at least one data buffer from which it unpacks data as needed by the peripheral device on the board. If the data needed by the peripheral is in the buffer, it is handed over. Otherwise the device is stalled while the interface performs a DVMA operation to fetch the requested data. Using address wrapping, the data that is needed shows up first, and the peripheral can be un-stalled while the rest of the data fills the buffer awaiting the next request.

Staying within block boundaries is helpful, too, because the aforementioned buffer can be simpler and can store fewer address bits with the data. This is especially important if there are multiple buffers, or if fast address-matching must be performed. With address wrapping, staying within block boundaries is automatic, and handled in hardware.

On the SBus, blocks start on addresses aligned to the transfer size. For 64-byte (16 word) transfers this means that only **PA(5:2)** change during the course of the transfer. The rest of the upper address bits are effectively the block address, and **PA(1:0)** must both equal 0 (again, because this is a word address). A good way of thinking about this is that these address bits are captured in a 4 bit binary counter which increments by one with each successive word transferred. Regardless of the starting address, once bits (5:2) count through 1111 (binary) they will wrap back to 0000 if the transfer has not yet completed, and continue on from there. A typical sequence here might be ??F0, ??F4, ??F8, ??FC, (wrapping back now) ??C0, ??C4, ??C8, ??CC, D0, ??D4, ??D8, ??DC, ??E0, ??E4, ??E8, ??EC.

For 32 byte transfers the same argument applies to **PA(4:2)**. For 16 bytes (4 words, as in our example in Figure 3.13.) the same is true for **PA(3:2)**, and for 8 bytes **PA(2)** is the only bit affected.

3.4.4 Consecutive (Atomic) Transfers

SBus atomic operations are tools which can help provide indivisible access to a register or memory location. Despite the name, these cannot guarantee true atomic access (see the discussion contained within Section 5.3.4), but can only guarantee exclusive access to the SBus. Future revisions of the SBus specification will call these *Consecutive Transfers* instead, but this book will continue to label them atomics for consistency with current designs and implementations.

This type of mechanism is useful when implementing resource locks. For example, a bus master wishing to gain exclusive access to the resource (such as a disk drive or a frame buffer) might need to test a status bit which indicates whether the resource is free. If it is, then the master will modify the bit, indicating that the resource is now in use. These operations must be consecutive, though, without possibility of interruption. Otherwise, two or more masters might each read that the resource is free and then try to gain control of it at the same time. Other bus architectures provide this kind of functionality via read-modify-write operations or locked transactions.

Normally, an SBus master de asserts its **BR*** signal for at least one clock after it receives a **BG***. If it does not, though, it will be guaranteed access to the bus both for this transfer and for the next. In this way the master can effectively lock the bus and guarantee itself exclusive access. A timing diagram which shows this is presented in Figure 3.14. Contrast this with the non-atomic example shown in Figure 3.10. Two or more transfers may be strung together using this mechanism.

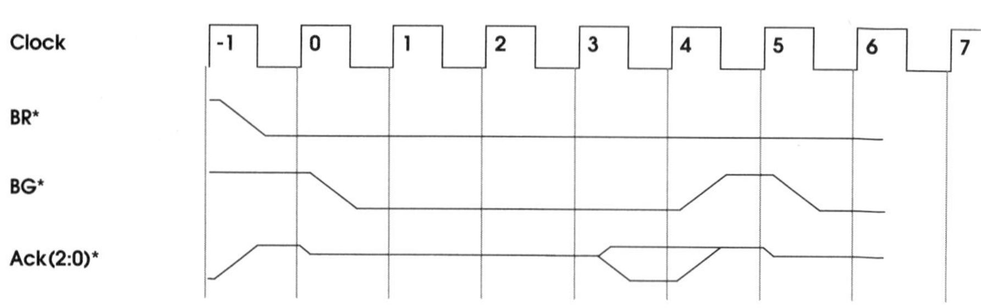

FIGURE 3.14. *An SBus Master Generates an Atomic Transfer by Holding BR*.*

This mechanism has a number of interesting implications. First, there is no up-front method for the controller to determine that an atomic operation is requested or in progress. Also, slaves and other masters cannot see this master's **BR*** signals, and so cannot determine that this is an atomic operation at any time during the transfer.

The use of atomic operations is heavily restricted by the specification. First, atomic operations should not be used to "cheat," and sidestep the normal arbitration process. For example, a master might seek to steal the bus' entire bandwidth for a short period of time by using atomics. This is inappropriate and unfair, and is likely to lead to system failure because the latencies for other devices will be substantially increased.

Another restriction on atomic operations concerns the order in which operations may be performed. The first operation must always be a read. The last operation must always be a write. These operations may use different addresses. In between, the master may stall for time if necessary by performing *dummy-read* transfers. These are read operations to the same address used on the initial read. One reason the master may do this is that while doing a read-modify-write operation, it requires extra time to perform the modification. Regardless of the reasons for a dummy-read transfer, it should only be considered a placeholder; the associated data may not be valid. Only the data from the initial read should actually be used.

There are many fine points involved in understanding and using SBus atomic operations properly. There are also some difficulties. Detailed discussions of these issues can be found in other parts of this book (see the Index) and are not replicated here. As a result of many of these issues, it is likely that future revisions of the SBus Specification now in progress will limit atomics even further, or eliminate them altogether.

SBus developers are discouraged from using atomic operations when an alternative exists. One possibility is that device drivers could arbitrate amongst themselves by using semaphores in structures contained within the system's main memory.

3.4.5 "Shadow" Register Operations

A special class of operations has been defined which can be used to access registers or memory not normally part of the device's address space. This is much like a secret door that can be opened when necessary, but is otherwise hidden from view. Many such doors can be built within any one slot, and each one opens up into a world that is exactly as deep as the slot's original address space. There are many ways in which this mechanism may be used. It will most commonly be used to increase a slot's physical address space, though.

For example, consider an expansion box that provides four new SBus slots in a "remote" chassis, which are all connected to the host via a special purpose board plugged into the "local" SBus. Somehow, one SBus slot's address space must be divided amongst four. This could be done evenly, or some slots could be given more space than others. Either way, though, the reduced address space available in each of the new slots may lead to undesirable configuration limits. Also, the expansion hardware may have its own registers and it must have an ID PROM. These could further reduce the address space available.

Rather than dividing the address space, though, it could be multiplexed. The expansion box might contain its own mapping hardware, which opens "windows" into larger address spaces. These windows can vary in number and in size, as long as the sum total size does not exceed the address space of the SBus slot at any one time. These windows can then be moved, opened, or closed as necessary. This is just one type of use which shadow register operations may have. For another, please see the description which starts on page 215).

Shadow register operations are triggered by writes to address offset 0 within a slot. Normally this space is the first location of the ID PROM, which is a read-only device. If a write occurs to this location, though, it has special meaning. Byte 0 (transferred on **D(31:24)**) contains a value which may be used as a "key." This is essentially a password which the slave verifies before accepting the operation. The key provides a level of security, but is also required to allow a bridge (which may use this mechanism) to pass on such operations to devices at the far end. The bridge hardware intercepts shadow register operations with the appropriate key, but is transparent to all others.

It is important that slaves do not hard-wire the key's value; this should be stored in a register which is written with the proper value when the device is initialized. Otherwise conflicts might occur with other devices, especially other devices of the same type. Slaves which do no support shadow register operations, or whose key values differ from that supplied, may ignore writes to address offset 0 or generate an error acknowledgment.

Even non-bridge slaves must qualify shadow register operations with keys. This might seem unnecessary at first. After all, the slave may be at the "end-of-the-line" (a leaf on the device tree). If the slave does not need to worry about passing these operations along, why must it make the extra effort required to obtain and use a key? Suppose this slave is on the remote side of some bridge device, though, that also uses shadow register operations. In this case, there is a danger that a write to address offset 0 might be intercepted by the bridge, even though it was directed at the remote slave. This would occur if the upper data byte of the write operation just happened to match the bridge's key. To eliminate this possibility, *all* slaves that make use of this feature must obtain and use their own unique keys.

When a write to address offset 0 occurs, the actual operation desired might be accomplished in any of a number of ways. At its simplest, it may toggle a bit which enables or disables a shadow register. That bit might also open or close an alternate address window. Or the write operation may contain data which indicates the type of operation to be performed. The mechanism is intentionally left flexible enough to allow a lot of leeway in its use.

3.4.6 Extended Mode (64-bit) Transfers

The SBus is a very simple interface at 32-bit widths, which was one of its primary design goals. Unlike many bus interfaces which multiplex address, data, and other information, the SBus separates these out in most cases into dedicated signals. This increases the number of signals required, but it also eliminates the need for address latches, reduces the opportunity for bus collisions (or the performance lost in trying to avoid them), and so on.

In systems which demand very high I/O bandwidths, performance can be more important than utter simplicity. Address and data multiplexing are not serious penalties if performance is boosted as a result.

The SBus' Extended Transfer Modes make use of this fact. The **Rd**, **Siz(2:0)**, and **PA(27:0)** signals are constant levels throughout a 32-bit wide transfer, but during an extended transfer these same signals carry different information at different points of the cycle. The **Rd** line indicates the transfer's direction for one clock only, then is driven low regardless of transfer direction. The **Siz** lines first indicate that an extended transfer is in progress (they do not indicate the transfer's actual size, though; more on that later). During the transfer phase these signals become **Data(63:32)**.

The SBus' extended transfer modes are more complex than their 32-bit counterparts, but don't require any more signals and are fully backward-compatible. Most importantly, up to twice the bandwidth is possible, because of a data path that is twice as wide.

Interestingly, extended mode transfers are also simpler than standard mode transfers in some respects! For example, bus-sizing is not supported, nor are slaves of various widths. There is no need; byte, half-word, or word only slaves can still be accessed using standard mode transfers. As a result, there is no need for byte steering.

Extended mode transfers are designed to be backward compatible with all standard mode designs. All interfaces which use the 64-bit protocols are required to also support the 32-bit protocols, so that transfers are possible at that level if all else fails. Existing designs which are not fully compliant with the specification may encounter problems, however. In order for a 32-bit SBus slave to be upward compatible and work properly in 64-bit SBus environments it is important that it fully decode the **Siz** bits, and respond with error acknowledgments for all transfer sizes not supported. Otherwise, it may respond to an Extended Mode encoding it did not account for in ways that the system cannot deal with.

A timing diagram for an extended transfer is shown in Figure 3.15. Compare this with the timing diagram for standard mode transfers, and notice that **BR***, **BG***, **AS***, **Sel***, **Ack(2:0)***, and **LateError*** are all used in exactly the same way!

Now look at the differences. Notice that the **PA** lines no longer transmit the physical address throughout the whole cycle. Instead, they carry the physical address (driven by the controller) for only one cycle, and then are tri-stated so that they may carry data (driven by either the master or the slave) later.

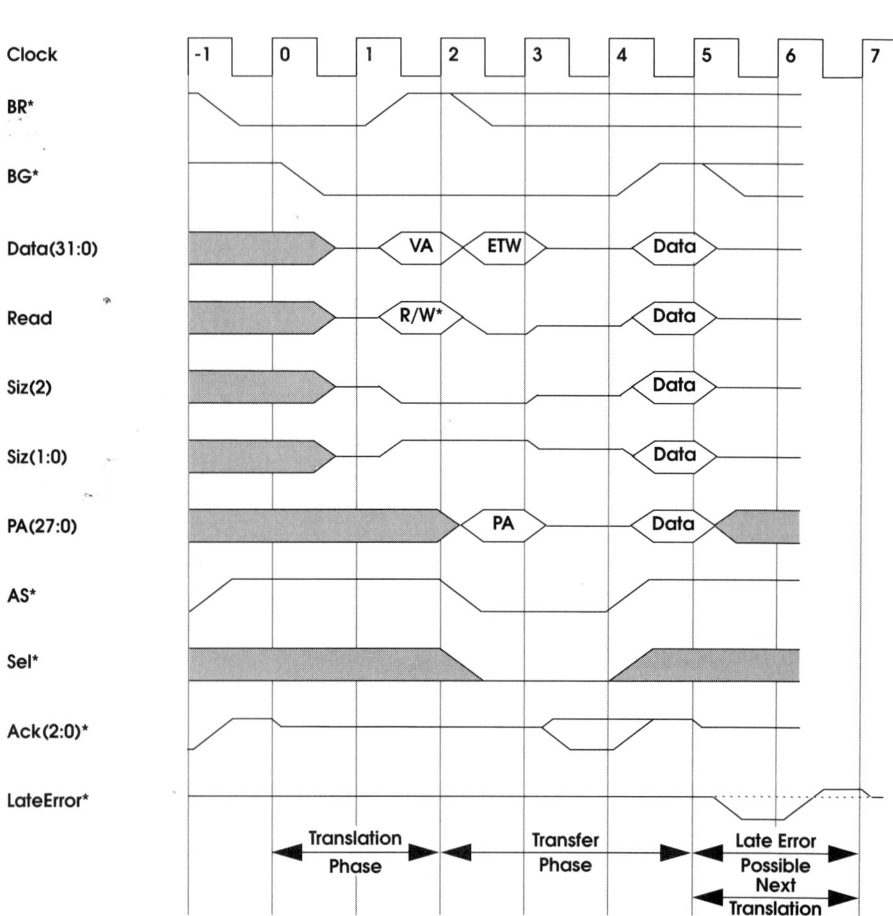

FIGURE 3.15. *Extended Mode Transfer.*

The behavior of the **Siz(2:0)** lines looks odd, but is actually straightforward. The master drives the extended transfer size code onto them in the early part of the cycle, just as some other size code would be driven early in a standard mode transfer. In this case **Siz(2)** is driven low, and **Siz(1:0)** are driven high. These levels must be maintained one clock past that in which **AS*** is asserted, so that the slave can sample them. The master must then tri-state these signals so that they can be used to carry data, but resistors hold **Siz(2)** down and **Siz(1:0)** up. This holds them in a state which continues to signal the extended transfer, yet can be easily over-driven when the time is right without worrying about bus-fights.

During extended mode transfers the **Rd** signal also behaves differently than it does during standard mode transfers. It only reflects the transfer's direction during the clock cycle in which the virtual address is on the **Data** bus. This allows the controller's memory management unit to qualify the translation (so that it can read- or write-protect parts of memory). After that, the master must drive the signal into the low state, whether or not a write or a read is in progress. It must then tri-state the signal, and a pull-down will hold it low until over driven later in the cycle by data.

This may seem counter-intuitive at first, but it helps ensure upward compatibility with existing SBus designs. As discussed in the previous section, all SBus devices should fully decode the **Siz** lines and behave accordingly. In case they do not, though, forcing the **Rd** line low should trick them into believing that a write operation is beginning. Hopefully, this prevents them from driving the data bus. This is not a bullet-proof solution, of course, but it limits the potential for incompatibility with existing less-than-rigorous designs.

Using the SBus' active-drive principle, resistors hold the **Siz** and **Rd** signals in certain states at some times during extended mode transfers, after drivers have actively moved the signals to the correct level. These states are vital to the correct operation of the transfer, and it is important that the values held are not corrupted. This could happen if the signals are glitched, or if the drivers begin to move them. This is similar to the holding amplifier metastability problem discussed in Section 5.1.2 on page 141.

The **D(31:0)** signals behave much as they did in the standard mode transfers. They still carry the virtual address to the controller, and they still carry data between the master and the slave. There are some differences, though. First note that the write data is not transferred before the acknowledgment in this mode, but after; just where the read data is transferred in both the standard and extended transfer modes. Also notice that there is no longer the herring-bone pattern which indicates permission to drive these lines before or after it is absolutely necessary. This is because those times are now needed to guarantee that there is no driver overlap. The biggest difference, though, is the presence of the *Extended Transfer Information Word* (abbreviated ETW). This is a descriptor provided by the master that contains detailed information about the type of transfer to be performed. The master must continue to drive this until it has sampled **AS*** and found it asserted. This allows the target slave to latch the ETW data when it finds **AS*** asserted, too.

The ETW contains several different fields, whose names and locations are shown in Figure 3.16. The first of these fields is the Extended Transfer Type. The encoding for this field is shown in Figure 3.17. Even though one of these codes is reserved (currently unused), it is important for developers to decode this bit and issue an error acknowledgment if the reserved function is indicated. This will help to assure compatibility should this code be assigned in the future.

Field Definition	Data Bits Used
ETType	D<31>
ETSize<2:0>	D<30:28>
ETRead	D<27>
ETAtomic<1:0>	D<26:25>
ETReserved<22:0>	D<24:0>

FIGURE 3.16. *Extended Transfer Information Word (ETW) Definition.*

ETType	Function
0	64-bit transfer
1	Reserved

FIGURE 3.17. *Extended Transfer Type Codes.*

At the start of an extended transfer, the **Siz(2:0)** signal field indicates only that an extended transfer is in progress. There were not enough unused size codes to allow the transfer's actual size to be encoded. Instead, that information is sent is encoded into the ETW's Extended Transfer type field, detailed in Figure 3.18. As previously discussed, all transfers are multiples of 8 bytes only. Again all bits should be decoded, and error acknowledgments issued for any size request not supported and on the reserved codes.

The ETW also contains information which declares atomic operations and identifies which part of the atomic sequence is in progress. The codes used are shown in Figure 3.19. This is a major advance over the method the standard transfer modes use to signal atomic transfers (see Section 3.4.4), because it provides more information, it occurs earlier in the cycle, and is visible to the slave.

ETSize(2:0)	Function
000	Reserved
001	Reserved
010	Reserved
011	One long-word (8 bytes)
100	Two long-word (16 bytes) burst
101	Four long-word (32 bytes) burst
110	Eight long-word (64 bytes) burst
111	Sixteen long-word (128 bytes) burst

FIGURE 3.18. *ExtendedTransferSize(2:0) Codes.*

ETAtomic(1:0)	Function
00	Non-atomic transfer
01	First atomic transfer
10	Intermediate (dummy) transfer
11	Last atomic transfer

FIGURE 3.19. *ExtendedTransferAtomic(2:0) Codes.*

The **DataParity** signal is not shown in the timing diagram, but its use also changes slightly during extended transfers. During the translation phase it provides odd parity for **D(31:0)**, as before. During the transfer phase, though, it provides odd parity for **D(63:0)**. This signal must be tri-stated whenever the data signals it covers are required to be tri-stated.

Extended mode burst transfers are no different than the single long-word transfer discussed here. As with standard mode transfers, burst transfers differ only in that more data is moved during each transfer. Extended mode burst transfers are also limited to at most 16 individual data "slices" per transfer, but because the data path is twice as wide the maximum burst size is increased from 64 to 128 bytes.

3.5 Performance and Latency

Sooner or later, everyone is interested in just what kind of performance you can get out of an I/O interface. Usually, raw data bandwidth is the figure most bandied about. Unfortunately, these figures are often relatively meaningless. The situation is not unlike that of knowing that a sports car can do 160 miles-per-hour when the legal speed limit is less than half of that. Unless bragging rights are the ultimate goal, other factors are far more important in measuring the performance of an I/O interface or an automobile. This section attempts to discuss a variety of performance issues, and how they are related.

3.5.1 Bandwidth

The SBus can transfer data 4 bytes per clock in the 32-bit modes and 8 bytes per clock in the 64-bit modes. For a 25 MHz clock rate this translates into a 100 Mbyte/second and a 200 Mbyte per second rate, respectively. Those interested only in raw (or *peak*) performance can skip now to the next section.

For the rest of us, the preceding numbers are not very meaningful because they take only a very narrow view of the transfer, and do not include the overhead required for each transfer. A more interesting number is the average (or *sustainable*) bandwidth. This is the bandwidth you could hope to see over an extended period of time, and includes all the intra- and inter-transaction overhead. For the SBus, there are at least four clocks of overhead associated with each transaction; one clock for address translation, one for asserting **AS***, one for de-asserting **AS*** (after the **Ack** has been received) and one for releasing the bus. To this we must add one clock per word (or long-word in the extended modes) transferred. A maximum length burst transfers 16 words (or long-words) and results in 20 clocks per transfer. This translates into an 80 Mbyte/second rate (or 160 Mbyte/second rate for extended mode transfers).

As with a car's mileage ratings, though, "your actual mileage may vary." While these rates are sustainable and could be seen over long periods of time, that is unlikely in a real-world application. That is because this assumes all transfers will occur back-to-back, and that all will use the most efficient transfer possible. In reality, both SBus controllers and slaves can insert wait-states at different parts of the transfer. This allows them to make trade-offs between cost and performance. Slaves of varying widths and bursts of various sizes can be supported for the same reason. Some transactions may not even transfer any data; they may be prematurely terminated with either error or rerun acknowledgments. The end result is that the actual bandwidth seen in a real mix of real components will be much less than the maximum possible. Just as the actual speeds you obtain with that fancy sports car will usually be limited by rush-hour traffic and your insurance company's tolerance for speeding tickets.

A graph which compares the SBus' optimum bandwidth versus burst size, and that of an actual SBus based host is shown in Figure 3.21. In this case the comparison is with a SPARCstation 1+, which was an early mainstay of the SBus host market. The graph also shows projected bandwidths for burst sizes not supported. This is to show the tendency for the bandwidth to asymptotically approach the bus' peak bandwidth as burst size increases. Compare this graph with the performance requirements summarized in Figure 1.1. This shows that the SBus bandwidth is more than adequate to support these and a wide variety other applications.

The graphs are drawn using data from Figure 3.20. This shows the number of *additional* clocks that the SPARCstation 1+ requires to perform certain operations. This machine requires a total of three clocks to perform address translations. It also requires two extra clocks for the first data word transferred, and two clocks per word thereafter. Notice that because ultimately two clocks per word transferred are required, this machine's peak data rate is 50 Mbytes/second. This performance is adequate for this machine, and its SBus interface is simplified by allowing additional clocks for some functions. This is important because the

SPARCstation 1+ is a low-end machine who's primary goal was to minimize the cost portion of the critical price/performance ratio.

Feature	Excess Clocks
Address translation Access time (1st word) Access time (subsequent words)	+2 clocks +2 clocks +1 clock
Burst Sizes supported:	4 word only (1 word = 4 bytes)

FIGURE 3.20. *Additional Transfer Overhead for the SPARCstation1/1+.*

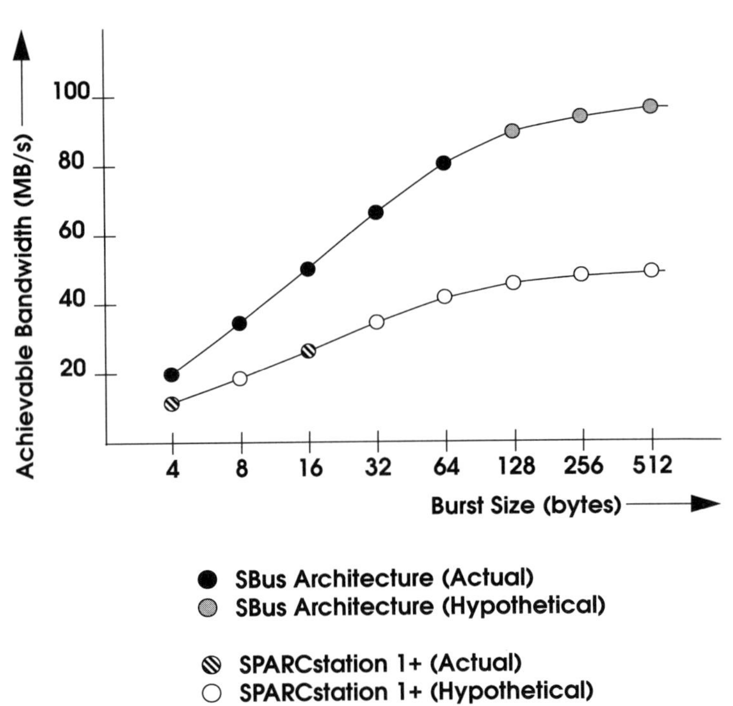

FIGURE 3.21. *Graph of Bandwidth vs. Burst Size (25 MHz writes).*

Other machines, of course, may choose the opposite strategy and optimize for high-performance instead of low cost. Faster internal clock rates, parts with lower access-times, and more parallel architectures are some of the methods that could be used to eliminate most or all additional overhead. Then the SBus' bandwidth would approach the architectural maximum. The SPARCstation 2 is just one example of a step in this direction. This machine's additional overhead is shown in Figure 3.22.

Feature	Excess Clocks
Address translation	+1 clocks
Access time (1st word)	+1 clocks
Access time (subsequent words)	+0 clocks
Burst Sizes supported:	all sizes

FIGURE 3.22. *Additional Transfer Overhead for the SPARCstation 2.*

3.5.2 Latency

The preceding discussion has shown that bus efficiency and bandwidth increase with burst size. It may seem desirable to support very large burst sizes, but in fact there are reasons why the burst sizes must be limited. The first reason is that not all devices transfer very large amounts of data at any one time. A few dozen bytes here and there is more common than large blocks. These devices couldn't readily use large burst transfers. Building data buffers into the device so that it does only transfer data in large chunks can be done in some cases, but in others this would add unwanted and unneeded design complexity.

More important, however, is the effect large burst sizes have on bus latency. Latency is a measure of the delay a device experiences between the time it first initiates (or requests) a data transfer, and the time the transfer actually completes. Excessive latency can cause device over-run or under-run errors, and it increases demands on the size and complexity of data buffers, FIFOs, and so on.

There are many factors that effect a bus' latency; among them are the number of requests that may already be waiting, and the time required to complete each request. This last factor is directly proportional to burst size, and is graphed in Figure 3.23. This graph shows the approximate latency (in milliseconds) for each

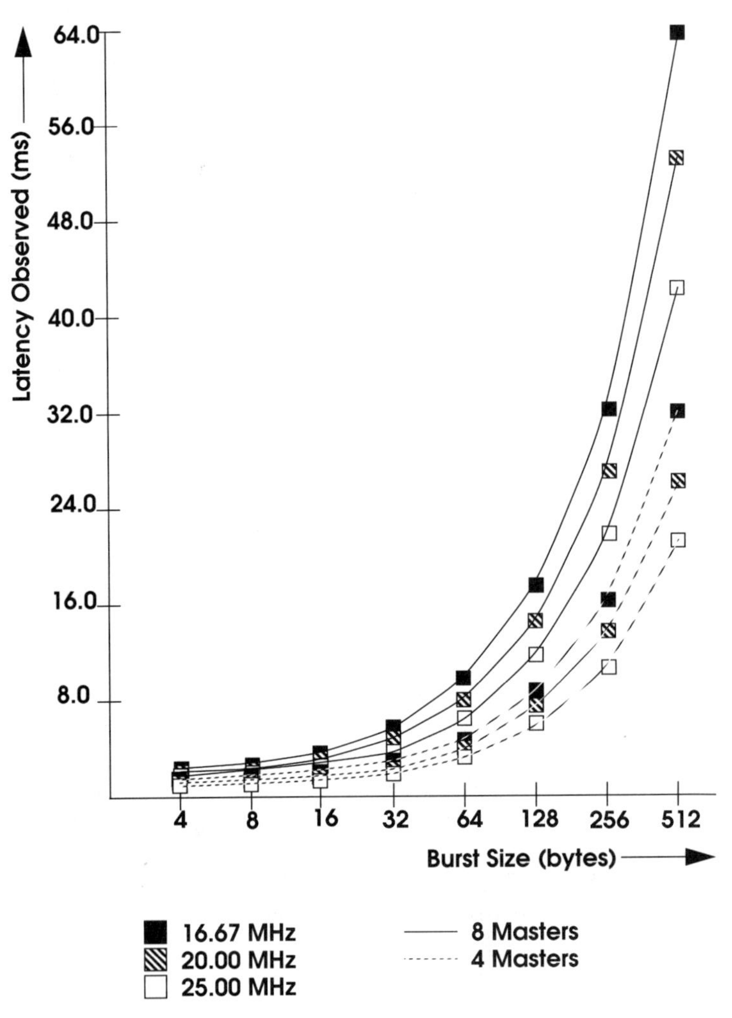

FIGURE 3.23. *Graph of Latency vs. Burst Size (25 MHz writes).*

burst size. These results assume a typical implementation in which the virtual address translation requires two clocks, the slave requires two clocks to begin responding, and then one clock per word thereafter.

The key result here is that unlike bandwidth, latency does not asymptotically approach some maximum as burst size increases. Instead, latency increases at an ever greater rate. This means that larger burst sizes produce diminishing bandwidth improvements at ever increasing latency penalties. This, too, is shown graphically, in Figure 3.24.

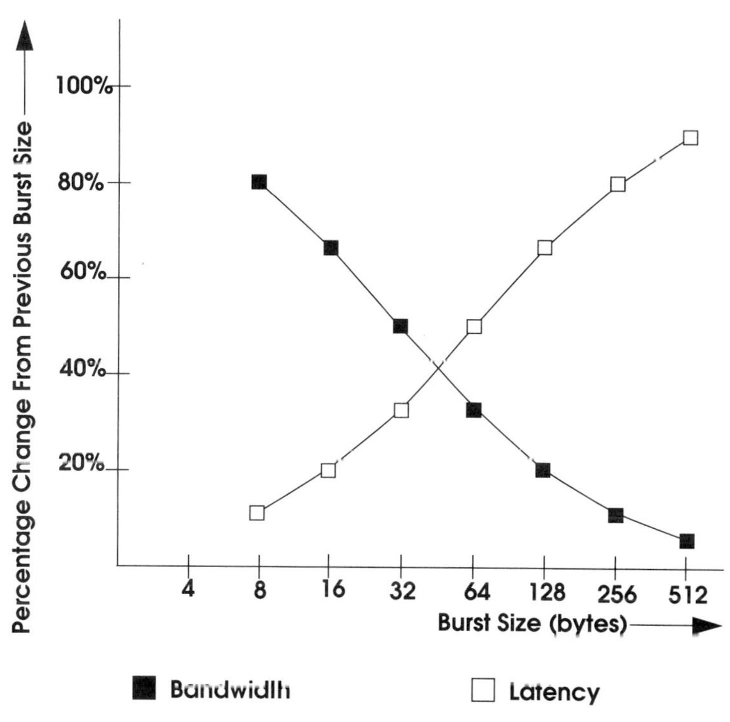

FIGURE 3.24. *Graphed Bandwidth and Latency Trade-offs (25 MHz writes).*

Here, the percentage bandwidth or latency change between burst sizes is shown. It is interesting to note that the percentage change is independent both of bus clock rate and of the number of masters on the bus. This is because these factors are constants which cancel. At the left side of the graph, increasing burst size brings with it a much greater bandwidth improvement than latency degradation. At the right side of the graph just the opposite is true, though. The crossover occurs toward the middle of the graph, and is dramatic when the step from 64 to 128 bytes is considered.

While improved efficiency and bandwidths are arguments for increasing the burst sizes supported by interfaces like SBus, the resulting latency increase provides a strong incentive to limit the upper bound. The 64 byte maximum SBus burst transfer size (128 bytes in 64-bit environments) seems a good trade-off.

SBus card designers should recognize that latency will be even greater on machines that must arbitrate for access to another bus. Interfaces like VME, Futurebus+, or MBus may be used at a higher architectural level than the SBus, for multiprocessor or processor-to-memory functions. A block diagram of such a system was shown in Figure 3.7 on page 51. In such a situation, access to two or more buses might need to be requested and granted before the transfer can complete. This will increase the latency observed.

3.5.3 Will Clock Frequencies Increase?

One question commonly asked about the SBus is whether or not its maximum clock rate will increase above 25 MHz. The best, simplest answer is that it will not; it doesn't need to, and to do so would result in a host of technical and compatibility problems.

For CMOS logic of the type used on the SBus, a 25 MHz SBus clock rate can be challenging. Above that frequency, the "pain threshold" increases rapidly. As clock periods diminish, rise and fall times, signal settling times, clock skew, and many other issues become far more critical, as does the inherent delay of the components themselves. Also, power consumption increases roughly proportional to the frequency.

The motive for an increased clock rate is extra bandwidth (which is directly proportional to clock rate). Some cards need very high bandwidths, but most do not (their bottlenecks are in the ethernet or SCSI interface, for example, not the SBus). Increasing the SBus clock rate to satisfy the needs of a few devices would penalize all the rest, because they are required to support the entire range of possible clock frequencies. This is directly contradictory to the SBus' ease of design goals. Many or most of the products in the installed base would not work at higher clock rates, and this might result in many configuration problems.

There are better, easier ways to increase bandwidths that do not require increasing the clock rate. The Extended Mode (64-bit) protocols are one good example; the 160 Mbyte/second rates that are possible with them far outstrip the demands of most current and projected I/O devices. Also, the simplicity of the SBus interface makes it fairly easy to replicate the bus on any given machine. I/O bandwidths can be linearly expanded in this way; a machine with two 64-bit SBus interfaces can move 320 Mbytes/second, for example! Better use of the bus' efficiency is another way to improve bandwidth. This means both better use of burst transfers, and less reliance on wait-states (during virtual address translation and data access).

In short, the SBus' maximum 25 MHz clock rate is unlikely to change because it is possible to achieve higher bandwidths through architectural, rather than technological, means.

Having just said all this, one caveat is necessary. The author is aware of many cases where a bus or a protocol or some other feature has far surpassed its expected life-span in the marketplace. It is possible that the SBus will live long enough that frequencies beyond 25 MHz are readily achievable. In that case, the upper limit of the SBus clock frequency may receive a mid-life rejuvenation at some point in the future.

3.6 Electrical Specification

The following sections discuss some of the SBus' primary electrical limits. The intent is not to tabulate all electrical specifications; that is the function of the specification itself. Instead, supplementary information is presented here to help understand the tables that are found in the specification.

3.6.1 Power Consumption

Single-width SBus cards may consume no more that 2 amps *average* current from the 5 volt supply. This average is measured over a 500 millisecond interval. Peak currents of up to 3 amps per slot may be consumed, as long the peak's duration does not exceed one millisecond, the average current is not exceeded, and adequate bypassing exists both in the host and on the SBus card.

The +12 volt and -12V supplies will not supply enough current for external ethernet multiplexors and the like. Their primary purpose is to provide bias voltages, or to power small amounts of analog circuitry such as op-amps, D/A converters, RS-232 converters, and so on. Single-width SBus cards may consume no more than 30 milliamps *average* from the +12 volt supply. The same limit applies to the -12V supply, which is independent. As before, the average is measured over a 500 millisecond interval. Peak currents of 50 milliamps are allowed, with the same restrictions specified for the 5 volt supply.

Double-width cards may use all the power available at each slot to which they are attached (double the single-width power).

3.6.2 Logic Levels, Leakage Currents, and Capacitance

The SBus' orientation toward CMOS technologies is a recurring theme throughout this book. There are several variants of CMOS logic, however, with differing input and output logic levels. The SBus uses that variant whose inputs and outputs both use "TTL compatible" logic levels. The capacitance and leakage current levels must be entirely CMOS compatible.

"TTL" Compatible Logic Levels

By definition, an SBus input is at a logic low level when at 0.8 volts (V_{IL}) or below, and at a logic high level when at 2.0 volts (V_{IH}) or above. To guarantee these levels, the output is expected to swing from below 0.4 volts (V_{OL}) to above 2.4 volts (V_{OH}). This is because there may be up to 0.4 volts lost as a result of output and trace impedances.

These voltage levels are roughly half of the more traditional CMOS voltage swings, which normally go from VDD (+5 volts) to VEE (ground). CMOS levels are a super-set of the voltages required by SBus, and so this type of driver could be used if necessary. Inputs that expect these voltages should not be used, though, because TTL compatible output drivers are not guaranteed to drive them properly. TTL compatible logic levels were chosen for the SBus because they are commonplace, and because they minimize the voltage swing. This last is important because the amplitude of signal reflections (hence overshoot and undershoot) is proportional to the signal swing, as is ground bounce. Also, the power dissipation into a capacitive load is proportional to the square of the voltage swing, so limiting the voltage swing reduces the power requirements.

CMOS Leakage Currents and Capacitance

While the logic *levels* are TTL compatible, that does not mean that the inputs and outputs themselves are. Ordinary (bipolar) TTL technologies should not be used to interface to the SBus. The capacitance and leakage currents of these inputs and outputs exceed those allowed by the SBus, sometimes by a large amount. The SBus capacitance and leakage current limits are very important to the bus' ability to drive signals at high rates despite low current capabilities.

For SBus add-on cards, the maximum leakage current allowed is 30 microamps DC. This means that the magnitude of the current measured flowing through the SBus connector pin must never exceed that value, over the entire range of input voltages. This applies to all SBus signals which are inputs to the card, or which are tri-stated outputs or transceivers. The capacitance must not exceed 20 pF total. This capacitance must include that of the connector pin (approximately 2 pF), all of the printed-circuit trace, all inputs and outputs attached to the net, and so on.

The **BR*** and **BG*** and **SEL*** signals are not bused; they are uniquely associated with each slot. Currently the leakage and current specifications for these signals is the same as for all other, bused, signals. That may be unnecessarily restrictive, and the limits for these particular signals may be increased in a future revision of the specification.

For SBus hosts there is no pre-defined leakage limit, although one may be added to a future revision of the specification. Prudence dictates that this should be minimized in any event. Total system capacitance is limited, however. This includes the capacitance of all SBus add-on cards installed, plus the connector, trace, and component pin capacitances within the host. This grand total sum must not exceed 160 pF for hosts with SBus clock rates of 20 MHz or less, or 100 pF for all other hosts. This applies to both bused and non-bused signals alike.

3.6.3 Signal Termination, Pull-ups, Pull-downs, and Holding Amplifiers

SBus signals are not terminated in the usual sense (see Section 7.1.1 on page 228), but certain signals must be either pulled-up, pulled-down, or otherwise held in valid states when un-driven.

This is because one time-honored axiom of digital hardware design is that unused or un-driven inputs should not be allowed to *float* (remain in a high-impedance state). This is especially critical for CMOS circuitry, which has very high input impedances. This makes it is very susceptible to false (unexpected) switching due to crosstalk, leakage, and other factors. This can cause serious problems if the signals serve a critical control function (such as **Ack(2:0)***). Even when not a direct problem, unnecessary switching can increase power consumption (CMOS power consumption is proportional to switching rate).

The selection of pull-up and pull-down resistor values involved many variables. The values must be high enough that standard ASIC output buffers can readily over-drive them. The values must be low enough that leakage currents can't over-drive them. As a result, the SBus uses 10 Kohm resistors for pull-ups, and 2 Kohm resistors for pull-downs. Pull-down resistors are stiffer than pullups for several reasons. The first is that many CMOS technologies contain internal pull-up resistors that must be overcome. Another reason is that some SBus card designs have incorrectly incorporated their own pull-up resistors, which also must be overcome.

The requirements for SBus pull-ups and pull-downs are summarized in Figure 3.25. It is the SBus controller's responsibility to provide these. Some signals do not require any kind of termination because they are always driven. The clock, **BG***, **AS***, and **Sel*** signals fall into this category.

For machines which do not support extended mode transfers, most of the physical address lines need no termination, either; they are always driven by the controller. In hosts capable of extended mode transfers, though, these lines can be tri-stated, and so need either pull-ups or holding amplifiers (see below). In either case, pull-down resistors should be used on any high-order address lines that are not driven by the controller. For example, machines which support only 25 physical address bits should pull-down **PA(27:25)**. This guarantees that such machines will correctly address the low-order address space in all cases, even with add-on boards capable of decoding all 28-bits.

Any host which supports the extended transfer modes must pull the **Rd** and **Siz(2)** signals low for reasons discussed in section 3.4.6. Machines which do not support the extended transfer modes may do this also or they may pull these signals high (or use a holding amplifier, described below), at their discretion. The **Ack(2:0)***, **LateError***, and **IRQ(7:1)*** signals must be pulled-high. The **BG*** signal must also be pulled high. It is not tri-stateable, but it is pos-sible that no card (or a slave-only card) will be installed in the asso-ciated slot. In that case, the pull-up is needed to guarantee that the signal is not asserted.

SBus Signal	Termination
Reset*, Clock	NA
Data<31:0>	10K pull-up/Holding Amp
DataParity	10K pull-up/Holding Amp
PA<27:0>	NA, 10K/Holding Amp[1]
BG*	NA
AS*, SEL*	NA
Rd, Siz<2>	2K pull-down/Holding Amp[2]
Siz<1:0>	10K pull-up/Holding Amp
BR*	10K pull-up
Ack<2:0>*	10K pull-up
LateError*	10K pull-up
IRQ<7:1>*	10K pull-up[3]

[1] Hosts which do not support extended mode transfers need no termination on any other address lines. Hosts which do support extended mode trans-fers should terminate all address lines used with 10K pull-ups or holding amplifiers. In either case, unused high-order address lines (such as PA<27:25> in a 25-bit address machine) should use a 2K pull-down.

[2] Hosts which do not support extended transfers may use a 10K pull-up (or a holding amplifier) instead.

[3] Future revisions of the SBus Specification may lower this value to as little as 2Kohm.

FIGURE 3.25. *SBus Signal Termination Requirements.*

The **Data(31:0)**, **DataParity**, and **Siz(1:0)** signals (**Siz(2:0)** in purely 32-bit hosts) are special cases. There is no real need to "de-assert" these signals because they are values, not levels. It is not necessary to move them to a logic high level. It is sufficient to guarantee that they remain at any valid logic level; high or low. In that case, *hold-ing amplifiers* are an alternative to pull-up resistors. A holding amplifier is essentially a latch that holds the line in its most recent valid state. This latch has an output impedance high enough so that it can be easily overdriven when it is necessary to impose a new state on the line. One possible configuration is shown in Figure 3.26.

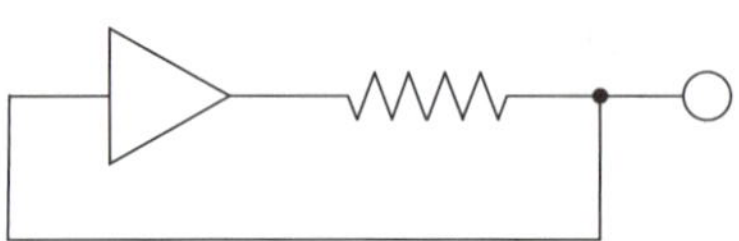

FIGURE 3.26. *One Possible Holding Amplifier Configuration.*

Holding amplifiers have two primary advantages over pull-up resistors. The first is that they are highly compatible with CMOS ASIC technologies and can be built right into the chip's input or output buffer cells. This can save parts; board space; and, ultimately, money.

The other advantage that holding amplifiers have over resistors is that they ultimately save power. First, a resistive pull-up burns power whenever it is overdriven. Additionally, if a line is always pulled high after it is driven then invariably it will switch more often than if it were just left where it was. Most power dissipation in CMOS circuitry is directly related to signal switching; the fewer times you switch it, the less power you use. Also, a signal that is pulled-up has a relatively long rise time. This allows it to linger in the receiver's high-gain threshold region where it is prone to oscillation. This results in even more switching and even more power dissipation. Remember, too, that all inputs and transceivers connected to the signal must be taken into account when considering the effects of extra edges and slow rise times. Multiply even a small power savings at each input by the number of signals involved and the number of inputs connected to each, and there is a substantial potential for total power savings.

Holding amplifiers do have a down-side that must be considered. Like any latch, they have setup and hold requirements, and may become metastable if these requirements are not satisfied. If the signal line is allowed to glitch, or if it begins to change state before the driver completely tri-states, then the holding amplifier may or may not hold it at a valid logic level. Since the holding amplifier is basically a feedback amplifier, it is likely to hold the signal at a mid-level for a relatively long time. This could dissipate even more power than a pull-up resistor would have in a similar case. For this reason it is strongly recommended that designers ensure that their outputs continue to maintain valid logic levels even while tri-stating. A more detailed discussion of this issue is contained in section 5.1.2.

Theoretically, holding amplifiers could be used on any of the tri-stateable signals, including **Ack(2:0)*** and **LateError***. After all, the principle of active drive requires that these signals be driven off before they are released; a holding amplifier could then hold the signal there. This is unwise in practice, however, because it places too much faith in the assumption that all participants are designed and functioning properly. Pull-up resistors are a much better choice for these signals because even if active drive is not implemented or working properly, the signals will still end up at the proper logic levels.

3.6.4 Timing

SBus' synchronous protocols greatly simplify its timing requirements. Compare the SBus with the asynchronous VME, for example. The SBus Specification summarizes the bus' timing requirements in one simple table. The VME Specification requires several tables and several timing diagrams to convey the equivalent information.

SBus timing is made even simpler because it has been reduced to two basic categories. The first of these is a set of expectations for the *driver* of any signal. The second set of these is a set of guarantees for the *receiver*. With very few exceptions these times do not vary from one signal to another; they are generic, independent of the bus' clock rate, and apply to all of SBus' synchronous signals. Also, all times share a common reference, and are measured from the rising-edge of the SBus clock.

Note that the driver (or source) of a signal does not necessarily mean an SBus master. Likewise, the receiver (or destination) does not necessarily mean an SBus slave. The driver is that which puts a value onto a signal, and the receiver is that which picks that value off the signal. For example, the master does drive the **DATA** lines during the virtual address translation and on writes, but it receives them on reads. It also receives the **ACK*** and **LateError*** signals, which may be driven by the slave. The SBus controller drives some signals (such as the **PA** lines) and receives others (such as **BR***).

There are several constraints placed on the driver of a signal. The most important of these are defined in Figure 3.27. First, the maximum amount of time available to drive a signal to a valid level is limited. This is called the Output Delay Time (abbreviated T_{OD}). The minimum amount of time that a signal must be held is also limited. This is called the Output Hold Time (abbreviated T_{OH}).

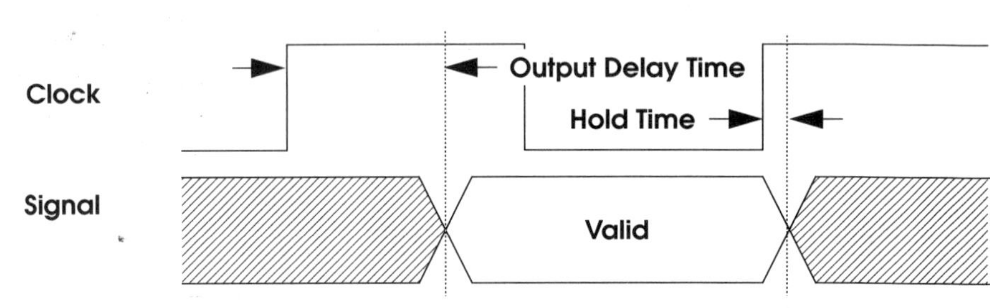

FIGURE 3.27. *Time Limits Imposed Upon the Driver.*

Revision B.0 of the SBus Specification limits T_{OD} to 22.5 ns in most cases. The only exception is for signals driven by a host whose SBus clock frequency is 20 MHz or less, in which case 32.5 ns is allowed. This exception specifically does not apply to add-ins, though, which must be designed to work across all possible frequencies. The minimum T_{OH} specified is 2.5 ns. This is the same value as the minimum T_{OD}, which is specified for consistency with the hold time limit. Please consult the specification for the conditions (capacitance and load resistance) under which these times apply.

The way that these times (and all others in this discussion) are measured is also specified. As mentioned, the rising-edge of the clock is used as the reference. Because the rise time of the clock is finite, the point midway between its V_{OH} and V_{OL} levels (or 1.4 volts) is specified as the exact point on which to base all measurements. The measurement's end-point is also specified as the midpoint in signal swing, this time of the signal under scrutiny. Measurements should be made at (or calculated with respect to) the connector at the signal's receiver. (Please see the discussion below for an explanation.)

If the signal's driver fulfills its obligations, then this provides the signal's receiver certain guarantees upon which it may base its timing. These guarantees are defined in Figure 3.28. The first is the input setup time (T_{IS}). This is the minimum amount of time that the receiver has before the clock edge to get the signal on-board, on-chip, and to any register or state-machine dependent on it. There is also a minimum Input Hold Time (T_{IH}).

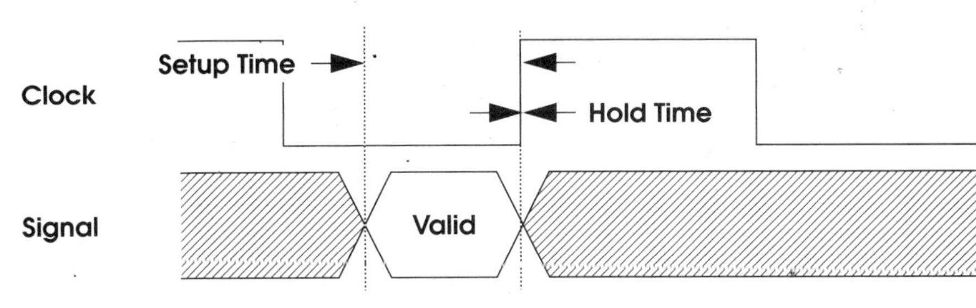

FIGURE 3.28. *Times Guaranteed at the Receiver.*

Revision B.0 of the SBus Specification guarantees T_{IS} will be at least 15 ns. The minimum T_{IH} will be at least 0 ns (in other words, the signal will never change before the clock edge). Again, please consult the specification for the exact conditions under which these times apply.

The relationship between the driver's requirements and the receiver's guarantees is close. The minimum SBus clock period (T_{CP}) = T_{OD}+T_{IS}+clock skew (T_{CS}). Similarly, T_{IH} = T_{OH}-T_{CS}. Clock skew is a factor not previously discussed. Due to finite propagation delays the rising-edge of the clock will not occur at exactly the same instant in time in all parts of the machine. There will be a difference, or error, or skew, when a comparison is made between the time the clock arrives at any two points in the system. By definition, the maximum clock skew equals the absolute value of the maximum such difference that can be found. The receiver's clock may happen *earlier* than the driver's, so this skew must be subtracted from the available setup time. Or the receiver's clock may happen *later* than the driver's, so this skew must be subtracted from the available hold time.

The SBus Specification limits clock skew to 2.5 ns or less. This will primarily be a function of the host's design, although the add-in card can adversely affect it if on-board clock trace lengths, capacitance, or buffer delays are excessive.

Notice that propagation delays have not yet been factored into these times. The specification doesn't fully address this, unfortunately. The specification only states that all times are specified "with respect to the SBus connector. Any additional times due to trace or logic delays [...] must be added or subtracted as appropriate." It is unclear from this, though, which connector is intended; the connector at the signal's driver, or at the receiver?

Ideally, the driver or receiver would only have to look as far as its own connector. Until propagation delays are specified though (and a corresponding constraint placed on system designers), this cannot be true, and either the driver or receiver must account for propagation delay (this will be labeled T_{PD} here). The driver is the best choice for this, because SBus timing is geared toward constraining the driver in order to provide guarantees to the receiver.

Until future revisions of the SBus Specification can clarify this issue, the connector at the receiver should serve as the basis for all timing analysis, and the driver must account for propagation delay. It is not sufficient for the driver to get a signal onto the bus within the T_{OD} limit; it must get the signal to the receiver's connector. To do this, it must add T_{PD} to its output delay time, and the sum must be less than T_{OD}.

What value should be used for T_{PD} though? Early SBus design work assumed a value of 2.5 ns. That is perhaps the best estimate available until a future revision of the SBus Specification can clarify the matter.

In addition to T_{OD} and T_{OH}, the specification also places a limit on the maximum amount of time after the clock edge which is available for tri-stating the signal. This limit (abbreviated T_Z) is defined as $(T_{CP}-5)$ ns. There is no corresponding minimum because it is ultimately limited by T_{OH}. The 5 ns is not arbitrary; it is intended to equal the sum $(T_{CS}+T_{PD})$. Interestingly, although T_{PD} is not defined in Rev. B.0 of the specification, it was assumed to be equal to the 2.5 ns previously suggested as an estimate. Note also that because add-in cards must work across the entire range of SBus clock frequencies, T_Z reduces to $(40-5)$, or 35, ns from their porspootivo.

The rise and fall times of signals are worth considering here, too. No signal rises instantaneously from the low to the high state, nor can it instantaneously fall from the high to the low state. A finite time is required to make this transition. This time is dependent on many factors, including the speed and power of the line's driver, the voltage swing, and the inductance, impedance and capacitance of the line itself. Whatever the factors, though, rise and fall times can add delay to signals because the signal won't actually pass the receiver's threshold for an additional period of time after it has started to rise or fall. The signal's driver must include these added delays in its T_{OD} measurements. That is the reason that this measurement and all others use 1.4 volts (midway between V_{OL} and V_{OH}) as the reference point.

With the driver given the responsibility for including rise and fall time delays in its time limits, it may seem oddly redundant that the specification includes maximum values for these delays. There is a good reason, however. If the rise or fall time is too slow then the signal may spend too much time in the receiver's threshold region. This could result in excess power dissipation or oscillation at the receiver.

Most signals have a rise or fall time limit of 20 ns. There are two exceptions, however. The clock signal must be more stringently controlled because it is such a critical time reference. Excess rise time (we're concerned mostly with the rising edge) would increase clock skew. Therefore the maximum rise or fall time of the clock is 3 ns.

The interrupt signals are the other exception. The fall time specification is 20 ns, just as it is for most other signals. These signals are open-collector, though; they are not actively de-asserted. Therefore the rise time may be very long, and the specification allows up to 1.2 μs! Future revisions of the specification may change this number somewhat, but the important point is that this is a long time. For this reason it is wise to use Schmidt-trigger receivers for these signals.

It may also seem odd that there are minimum rise and fall time limits, too. After all, it seems logical that to minimize signal delays very fast rise times are desirable. That is true, of course, but this is one good example where it is possible to have too much of a good thing. Very fast rise times increase the degree to which the signal will behave like a transmission line. This is undesirable in the SBus' case because it can lead to excessive signal ringing, overshoot, and undershoot (for more information, see section 7.1.1).

The minimum rise or fall time of the clock is 1 nanosecond. For all other signals revision B.0 of the specification sets the minimum at 5 ns. This last number will likely change in future revisions of the specification because it is very difficult to guarantee with the technologies most commonly used on the SBus (see section 5.2.4).

Rise and fall times are measured between the 10% and 90% point of the signal's swing.

3.6.5 Stub Length Limits

The traces on SBus cards which take signals from the connector to the drivers or receivers must be limited in total length. In part this is to limit capacitance, but mostly it is to limit transmission-line *stub* effects. A stub is a branch off of the signal's main path, as shown in Figure 3.29. Stubs complicate the line's impedance and can lump excess capacitance at the point of connection. All this degrades the signal line characteristics by fostering signal reflections which result in ringing, overshoot, and undershoot. If the stub is excessively long it can even behave like a transmission line in its own right, and can add to signal reflections in this way. More information on transmission-lines and signal reflections can be found in section 7.1.1.

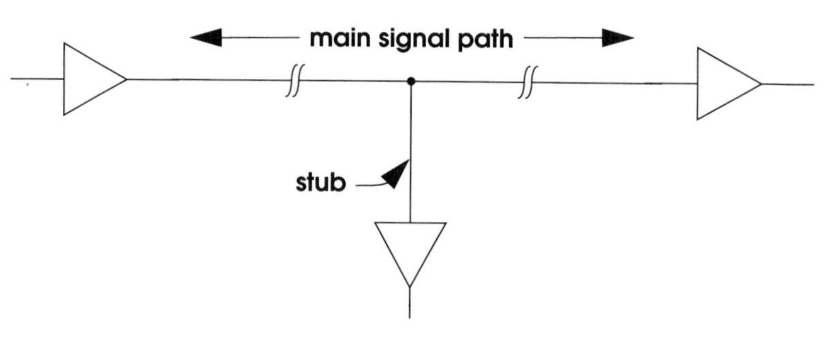

FIGURE 3.29. *Stubs are Branches off the Signal's Main Path.*

The SBus Specification limits stub-lengths to at most 50.8 mm (2"). Revision B.0 applies this limit to all signals, but future revisions will allow an exception for the **IRQ(7:1)*** and **Reset*** signals. Routing of these can be more flexible, and so stub lengths of up to 76.2 mm (3") will be acceptable.

What of the case where a trace runs from the SBus connector to multiple loads? Such a trace may itself contain branches. Does the stub length limit then apply to the sum of all trace lengths, or the maximum path length to any given end-point? The specification does not define this, unfortunately. The most conservative approach, though, is to run such traces sequentially from the connector to each load, without branches or internal stubs, keeping the total length less than the stub-length limit. Capacitance limits will make it difficult to have multiple loads and multiple traces in any event.

The specification also does not define how or if the stub-length limits apply to traces within the host. However, it makes sense to also route these signals sequentially when possible. When branches or stubs are necessary, keep them within the aforementioned limits.

3.7 Mechanical Specification

The following sections focus on mechanical issues. As with this book's discussion of the SBus electrical specification, this material is intended to supplement the specification, not replace it.

3.7.1 Board Size

The SBus Specification defines two different board sizes, which occupy one or two SBus slots. A third size also exists which occupies three SBus slots, but this form is not defined by the specification and is not encouraged.

Single-Width

Most SBus cards use the *single-width* form factor, so called because only a single SBus slot is occupied by the card. The key dimensions of this card are diagrammed in Figure 3.30. The board is very small; only slightly bigger than a regular 3" by 5" index card.

The small size of the board is a result of SBus' emphasis on highly integrated applications. While the board area is small, so are the ASICs and surface-mount components for which the interface is designed. Also, the SBus interface is simple, generally does not require buffers, and so usually requires very little area, leaving the bulk of the board's area to the add-on's intended function.

In addition, space is reserved for components on both the top and the bottom of the board, effectively doubling the area available. The height below the card is limited, but is ample for both active and passive surface-mount components.

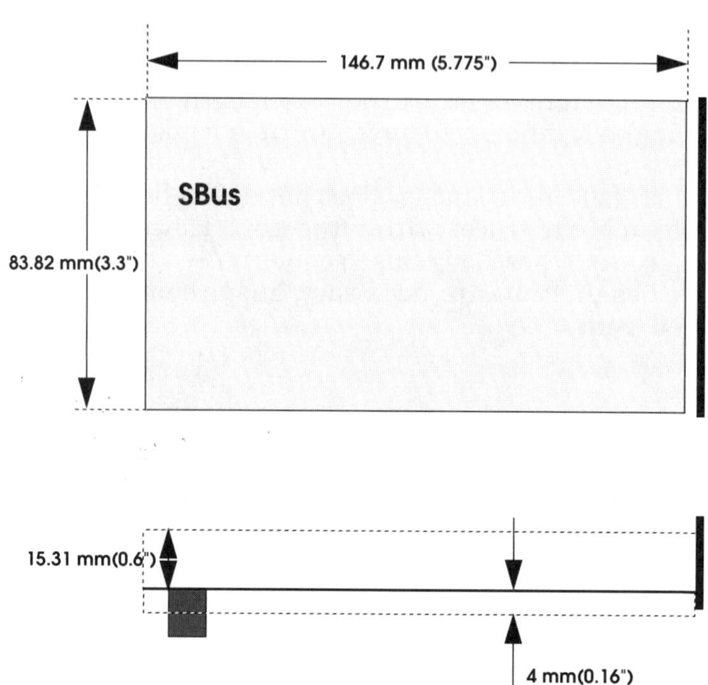

FIGURE 3.30. *SBus Single-Width Card Dimensions.*

Double-Width

If a single-width SBus card does not provide enough power or board area, then a double-width card can be built. As the name implies, these boards are twice the width of a single-width card, or 170.28 mm (3.7"). These cards have two SBus connectors, and are designed so that they plug into two adjacent slots in the host.

Double-width cards may use all the power allocated to each slot in which they are connected. This allows them up to twice the power budget of a single-width card, and is one primary reason that double-width cards are built. Regardless of their power requirements, though, a double-width card must draw its power as evenly as possible from both slots. Just as all the power pins in any one slot must be connected, so too must all the power pins in both connectors of a double-width card. In both cases this requirement is intended to minimize impedance in the power distribution system.

All SBus signals, however, should only be connected to one slot or the other. This may seem an unusual and unnecessary restriction, but there are at least two good reasons for it. The primary intent is to limit differential loading, which could introduce additional skew into signals. The more important reason, however, is a result of the multi-channel architectures discussed in Section 3.3.3 above. In such an architecture, it's possible that two adjacent slots may not both be connected to the same SBus! If all signals are drawn from one connector or the other, the board will work fine in such a situation. If some signals are drawn from one connector and the rest from the other, however, then the board may find itself straddling two buses and unable to communicate with either.

Double-width cards should only be used as a last resort because their size restricts their use. There are several SBus hosts which only have two slots altogether. A double-width card would completely fill them and leave no room for expansion. Not even a bus-bridge or expansion box could be used to remedy the situation! Even worse, some such hosts require the use of an SBus frame-buffer, which leaves only one slot free. Others have only one slot to start with. A double-width card is not compatible with either of these.

Triple-Width

Triple-width cards have been built. These occupy three slots total. The SBus Specification does not describe triple-width cards, and instead actively discourages them. The circumstances that make a double-width card undesirable become acute when triple-width cards are considered. Also, while many systems existing today do have the three adjacent, evenly spaced slots that make triple-width cards possible, there is no guarantee that this will always be the case.

Volume

The SBus card may occupy the entire volume of space above and below the board area, up to the height limits. The area above the SBus connector is included in this, as is the area below the card near the backplate.

There can be confusion here because there are some drawings in SBus literature which have dashed outlines that at first seem to indicate "forbidden" zones. Closer inspection reveals that there are no dimensions associated with these, however. That is because there are no forbidden zones, and the dashed outlines shown in some of the specification's profile drawings are relics left over from earlier revisions. Schedule constraints or simple oversight have prevented them from being fine-tuned.

Minimum Below-Board Gap

Developers building an SBus host must guarantee that a gap of at least 1.52 mm (.060") exists between all components on the solder-side of an SBus card, and between all components on the mother-board itself. This gap is required to prevent any unintentional contact between the add-in and the motherboard, even under extremes of board warpage and mechanical shock and vibration.

This issue does not affect add-in card developers as long as they remain within the 4 millimeter solder-side component height limitation.

3.7.2 Connectors

The SBus uses 96 pin "mini-din" expansion connectors. These are high-density connectors that have two rows of 48 pins each. The rows are spaced 2.54 mm (.100") apart, and the pins themselves are on 1.27 mm (.050") centers. The male connector is mounted to the solder-side of the SBus add-on card, and the female connector is attached to the component-side of the host motherboard. Please refer to the SBus Specification for detailed mechanical outlines of this connector, and for lists of vendors which can supply them.

Some versions of these connectors are equipped with a "key." This is a plastic tab or pin that is on one side of the connector only, and is designed to fit into a specially prepared hole in the printed-circuit board. Its purpose is to prevent the connector from being inserted incorrectly in the board. On any attempt to do so the pin will not allow the connector to seat properly. Note that if the connector pattern used includes the keying hole, either keyed or non-keyed connectors may be inserted. This is a wise precaution because it maximizes the vendor's flexibility by increasing the number of choices available.

Surface mount versions of the connectors are available and may be used if desired. If so, strain relief should be provided by fastening the connector to the PC board using the tabs provided for this purpose. Through-hole connectors need no such relief because the pins themselves provide sufficient mechanical strength. Hosts may vary the height of the connector, as long as corresponding adjustments are made in the backpanel or frontpanel, and any other place necessary.

3.7.3 Backplates

SBus cards must contain a backplate, which is a mechanical bulkhead usually made of metal. Designed to mate with backpanel or front panel of an SBus host, the backplate and the SBus connector form the primary mechanical support of the board in the host. Access from the "outside world" to electrical connectors on the board may be gained through the backplate. The backplate is also intended to complete the EMI enclosure of the host, plugging and shielding the connector cutouts in its backpanel or frontpanel. A diagram of a typical backplate is shown in Figure 3.31.

Notice that the backplate is composed of two pieces. The upper thin piece provides the mechanical tabs (or "ears") usually necessary in desktop, user-installable, "no-tools-required" installations. It is removable for those applications where board height must be minimized. See Section 3.7.5 below for more information.

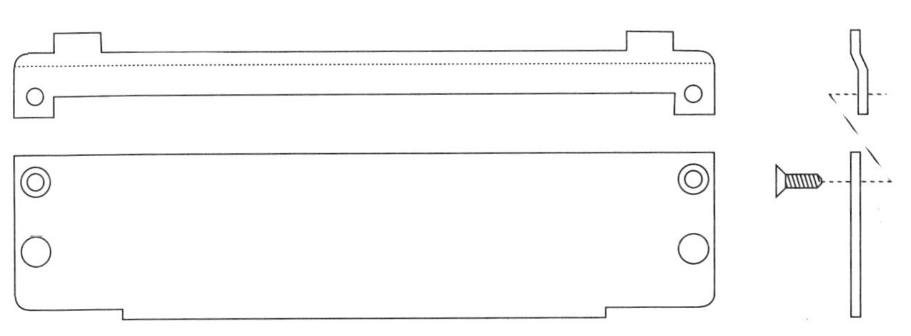

FIGURE 3.31. *Single-width SBus backplate.*

Double-width cards may use one double-width backplate, as mentioned in the SBus Specification. They may also use two single-width backplates, and this is often the better alternative because it increases the chance of finding the backplates you need "off-the-shelf."

Under no circumstances should the backplate be connected to logic ground. For a complete discussion of the reasons for this restriction and the alternatives, see the discussion in Section 7.2.2.

Do not use a single-piece backplate under any circumstances. While this might seem simpler and less expensive in the short-term, it will cause increasing compatibility problems in the future as more and more SBus hosts require low-profile SBus cards.

Connector "Tunnel"

Many SBus devices require connections to devices outside of the host's enclosure. For example, a frame buffer must be connected to a CRT, and an ethernet card must be connected to the network. Connectors are used, of course, and these must pass through the backplate and then through the backpanel or frontpanel to get to the "outside world." This necessity forces some size constraints on these connectors.

The backplate drawings in the SBus Specification contain a dashed outline which defines the *tunnel* which constrains these connectors. This is called a tunnel because it can best be visualized as a rectangular tunnel which projects both in front of and behind the backplate. Any connector used on the SBus card must fit entirely within this tunnel. This includes all mounting screws, brackets, and so on. It also includes the cable which will be attached to the connector. This last restriction is necessary because the backpanel may be relatively thick, and the SBus card connectors will be recessed within them.

Revision B.0 of the specification does not limit the distance that a connector may extend beyond the backplate, as long as it stays within its tunnel. Future revisions of the specification may be forced to limit this, however. That is because very long extensions complicate the task of installing the SBus card. By design, the card must be angled into place before the SBus connectors can be fully mated. If the connector extends too far beyond the backpanel then this can make it difficult or impossible to maneuver the card into place in some hosts.

Due to system grounding concerns, it may be necessary to electrically isolate some connectors from the backplate through which they pass. For more information, see section 7.2.2.

3.7.4 Use in "Desktop" Environments

In desktop installations ease of use is a key requirement, and installation can be a big part of this. The SBus is designed to allow installation without tools in desktop environments. Cards may be simply inserted and pressed into place. The card's backplate must have the upper piece (with the mechanical tabs) attached to it.

Mechanical support is also provided by the SBus connector, but a plastic retainer is used to keep the card from wiggling its way out of the connector because of shock or vibration.

One possible design of the SBus retainer is shown in Figure 3.32. The retainer looks like a handle, and that is what is was originally called, in fact. That is not its primary purpose, though, and it may break if too much force is applied. Its primary purpose is to keep the SBus card firmly seated in its connector. The top of the retainer is meant to make contact with the inner surface of the enclosure's cover. In this way, any shock or vibrational force that might have worked the board out of the connector is transferred to the enclosure instead.

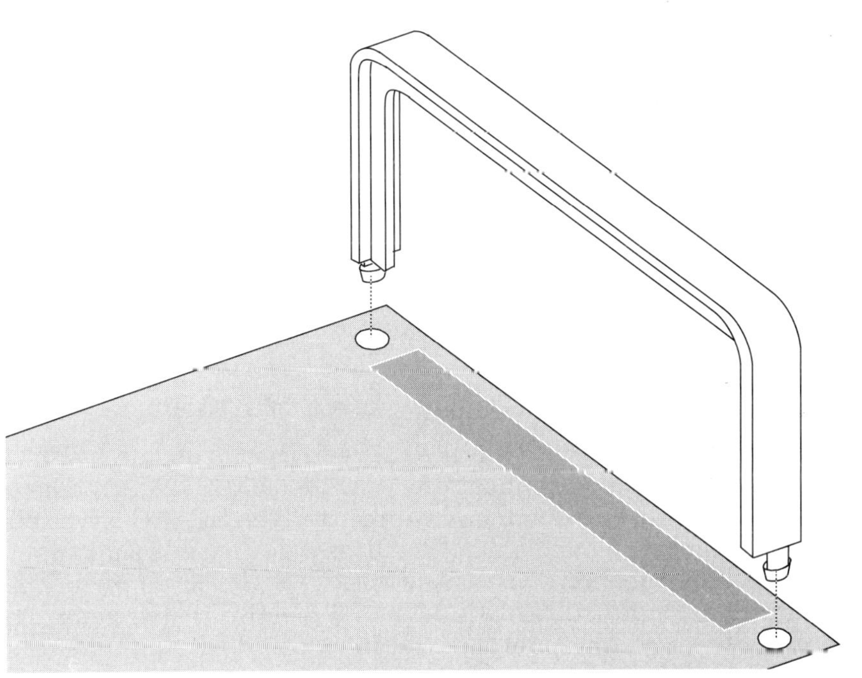

FIGURE 3.32. *One Possible Retainer Design.*

The retainer is designed to snap onto the SBus card. One clip on each side fits into a corresponding hole on the card on either side of the SBus connector. This arrangement allows it to be easily installed or removed, as necessary.

Other retainer designs are possible. That shown in figure Figure 3.32. is the one published in Revision B.0 of the specification, and the impression left is that it is the only "approved" design. This was not the intention, however; this design was meant to be typical, not mandatory. Future revisions of the SBus Specification (now in progress) allow developers the freedom to customize the retainer design as long as the changes made do not adversely affect the fit or function. This will be especially useful to host developers who may need a special retainer or mechanism for their particular enclosure.

SBus cards must be shipped with a "typical" retainer (primarily, this means the height and the placement of the holes used to attach the retainer to the board must correspond to those published). If such a retainer will not work in a particular host, it is the responsibility of the host to provide an alternative retention mechanism.

While not mandatory, use of the published retainer designs provides several advantages. A common "look-and-feel" is one result. Procurement costs may also be reduced if a limited subset of designs are available as commodities "off-the-shelf" from commonly used distributors. Also, the chances for future compatibility problems are reduced because the number of possible retainers (and hence configurations) is reduced.

3.7.5 Use in "Backplane" Environments

Some SBus hosts place additional restrictions on the height of an SBus card. These applications include VME, Futurebus+, and other backplane oriented environments. The SPARCserver 600MP is one example of such a machine. It is also possible that some very high density desktop hosts may "stack" SBus cards in such a way that would require these additional height restrictions. Highly integrated, or "laptop" style machines would be another example.

For environments such as this, two changes must be made to an SBus card. First, the upper piece of the backplate must be removed. The screws should be saved because they will be used to fasten the backplate to the host's backpanel or frontpanel. Second, the retainer should be un-clipped and removed. In its place, some form of stand-offs (also specified in the SBus Specification) will be used to provide retention in case of shock and vibration.

SBus stand-offs are actually a special form of retainer, optimized for low-profile environments. One example of such a stand-off is shown in Figure 3.33. As with retainers, this design is only typical of that which can be used. The stand-offs actually used might vary, especially in length. Designs which allow the card to be snapped in and out are preferable, but even this is not required as long as the mounting holes specified for retainers can be used as is.

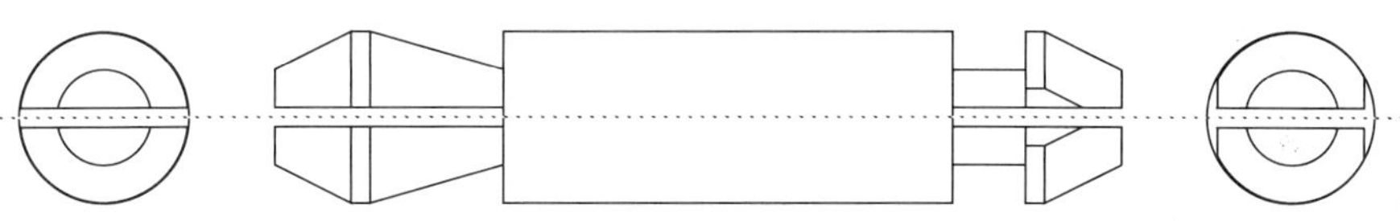

FIGURE 3.33. *SBus Stand-offs Used in Low-Profile Applications.*

If stand-offs are necessary, they must always be provided by the host machine.

3.7.6 Temperature Range

SBus products must be designed to work at ambient temperatures between 0 and 70 degrees centigrade. This is the temperature measured at the slot itself.

An SBus card may dissipate up to 10.7 watts total per slot (based on the power consumption limits discussed previously). Host designers must plan the system's packaging and cooling to guarantee that this heat is adequately dissipated and that the specified temperature conditions are not exceeded. This does not necessarily require forced air, and some hosts (especially lap-tops) might rely on convection cooling.

The SBus card should therefore be designed with components that will work properly across this entire temperature range. As just mentioned, no airflow can be guaranteed; still-air must be assumed. Beyond this requirement, SBus cards should also be designed with the problem of heat removal in mind.

For example, some systems do provide air flow. This will almost certainly be in a direction perpendicular to the add-on's long axis (see Figure 3.34), because otherwise it would be blocked by the SBus connector, backplate, and backpanel. Still, even in the preferred direction the airflow can be blocked by components on the board, or on another SBus board "upstream." Components placed perpendicular to the flow may block most of the air moving past them, especially if they are tall or are socketed. It is much better to mount components parallel to the slipstream when possible. Try to use low-profile parts; surface mount devices are especially nice in this respect. Also, eliminate sockets which aren't absolutely necessary. Any thing that can be done to minimize the cross-sectional area of the board will reduce its airflow restriction.

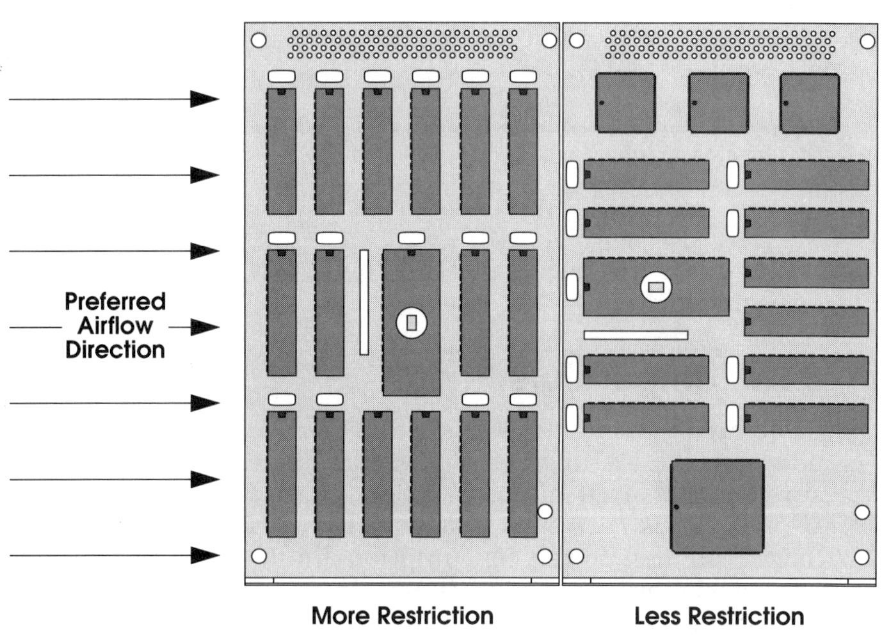

FIGURE 3.34. *Limit Perpendicular Airflow Restrictions.*

Another issue is the removal of heat generated by active components mounted on the solder-side of an SBus board. Any power dissipated by those components is trapped between the card and the motherboard along the vertical axis, and by the SBus connector and backpanel along the card's long axis. This is a narrow, cramped space which can be heavily constricted by components on the card and on the motherboard. Removal of heat from this area can be difficult, even with forced air. It is therefore wise to limit the amount of heat generated there, and the SBus Specification recommends a 2 watt limit per slot.

3.7.7 Board Thickness

The specified thickness for an SBus card is 1.60 millimeters (± .20 mm).

If the number of board layers your design requires will not fit into this dimension, it's understandably tempting to consider increasing the board thickness a small amount. After all, the SBus does not require the use of card guides which might bind or stick on the card if it were too thick. There are other factors, though, which make it inadvisable to increase the thickness of the SBus card.

One factor is that the component height is measured from the bottom surface of the board. Any additional thickness will subtract directly from that available for components. This in itself may not be a big problem, but it is something to consider.

Other problems relate to the SBus retainer and stand-off. Both the stand-off and retainer must fasten securely through two holes in the board. If the board is too thick they will not grip properly and may even break.

Also, if the board's backplate is fastened to the top surface of the board in any way (or if it is attached to a connector which is fastened to the top surface) then it will be too high if the board is too thick. This would cause the card to fit only at an angle, which could interfere with the SBus connector, the backplate's connection to the backpanel, or with component clearance on the solder-side of the board.

Acknowledgments

Figure 3.12. is reprinted with permission of Troubador Technologies.

References

Ott, H. W. *Noise Reduction Techniques In Electronic Systems*. John Wiley and Sons, 1976.

Lyle, J., M. Sodos, and S. Carrie, "The SBus: Designed and Optimized for the User, the Developer, and the Manager." *BUSCON-East Conference Proceedings*, October 1990.

Turner, R. P. and Gibilisco, S. *The Illustrated Dictionary of Electronics.* TAB Books, 1988.

Firmware Concepts

4

All SBus cards must contain an ID PROM that minimally contains a " name" attribute. The ID PROM may optionally contain a number of other attributes, as well as device-specific firmware that can help to install or diagnose the card. The structure of the attributes and code contained within the ID PROM is defined by the Open Boot Architecture, which was developed in conjunction with the SBus.

The SBus Specification does not require support of the Open Boot Architecture. Only the presence of the ID PROM and the name attribute are essential on an SBus card, and all other features may be ignored. An SBus-based host may ignore even the ID PROM if desired, letting the device's driver assume the responsibilities that the ID PROM might otherwise have managed. The card designer may not use this as an excuse to eliminate the ID PROM, however.

Though not required, the Open Boot Architecture provides significant benefits. Chief among these is the ability to easily integrate new boot-devices. Most SBus-based hosts will support some variant of the Open Boot Architecture.

4.1 The Open Boot Architecture

The firmware in any host is responsible for controlling the machine between the time it first starts to operate and the time it is capable of loading and executing the operating system. Firmware usually includes code that initializes the processor(s), tests memory, and performs other diagnostics. It must also contain at least rudimentary device drivers for whatever devices must be used to load the operating system. If the operating system resides on a local hard disk, then a driver for that is required. If the operating system resides on a server reached through an ethernet interface, then an ethernet driver is required. Basic CRT, keyboard, and other drivers are usually also necessary so that error messages can be reported or choices about the boot process can be made.

The firmware PROMs are an integral part of the system when first shipped. They are not easily or commonly changed in the field. This often makes it impossible for the built-in firmware to keep up with I/O device technology, which may change dramatically over the course of the host's lifetime. Also, there might be a wide variety of devices available from a wide variety of vendors; supporting even the most common configurations could rapidly become unwieldy.

4.1.1 Background

A history lesson from the IBM-compatible personal computer market is enlightening here. At power-up the operation of these machines is controlled by the Basic Input/Output System (BIOS) contained in PROM on the motherboard. When these machines were first introduced in the early 1980's, the primary mass storage device was a cassette-tape interface. This was followed first by a single-sided 5.25 inch floppy diskette, then by a double-sided diskette, then by a double-sided, double density diskette, and then 3.5 inch diskettes came along. Soon hard-disks also became standard equipment. Originally 10 Megabyte disks were considered large, but now disks with even hundreds of Megabytes are commonplace. The operating system and application programs, though, might not even reside on a local disk anymore; they might be on a file server at the other end of an ethernet or FDDI interface! Undoubtedly, BIOS's original cassette interface driver is not at all suited for use with an ethernet port.

Obsolescence helps, of course. The BIOS PROMs found on PCs bought today are significantly more sophisticated than those on the original machines. But it's not practical to throw out the whole machine whenever you need a bigger disk or a new network interface. Nor is it practical to introduce a new product and then wait patiently for the installed base to evolve to a point where it can make use of it in volumes that are worthwhile. Some mechanism is needed to allow the system's firmware to keep up with new developments.

Shipping a replacement set of PROMs with your product isn't feasible either. Even machines that are compatible with each other are usually not identical. Different PROMs would be necessary for different machines. Besides, changing PROMs when installing boards is yet another hassle for end-users, and yet another chance for them to make mistakes that will result in angry phone calls or returned products. Further, what do end users do when they get one PROM for the disk controller, another for the CRT interface, and a third for the ethernet port? They won't all fit into the same socket!

It is possible to ship a PROM *extension* with your product, however. A simple interface can be built into the system's original firmware that allows it to recognize and execute additional firmware on a new device. This is the mechanism used within the IBM-compatible market that allows the computer's firmware to keep up with the latest state of the art.

4.1.2 Strategy and Requirements

A mechanism much like this is also used within the SBus realm. The Open Boot Architecture provides a basic set of functions. More importantly, however, it also provides a well defined interface that allows additional functions and features to be tightly interlocked with the original feature set, as shown in Figure 4.1. This allows the host to use (and boot using) devices that might not even have been conceived when the host was first shipped to the customer. These additional functions and features are contained within an ID PROM resident on each SBus card.

The Open Boot PROM Architecture must be more sophisticated than the related mechanism used in the IBM-compatible PC realm, however. The PC realm is based on Intel's 8086 microprocessor family. The instruction set is standardized and the BIOS structure has been optimized around it. Also, there is no built-in support for autoconfiguration. It is often necessary for the person installing a device in this environment to set DIP-switches and install jumpers and modify configuration files. As a result, it's also quite possible for one device to conflict with another, and great care must be taken when installing and configuring a new device.

The SBus is architecturally independent, though. It can be SPARC based, or 8086 based, or 68000 based, etc. The firmware extensions must therefore be written without making any assumptions about the base instruction set. Also, autoconfiguration is an important SBus goal. It should not be necessary to set DIP-switches or install jumpers, and the firmware must consequently take responsibility for configuring the system and avoiding conflicts. These requirements factor heavily into the resulting structure of the Open Boot PROM and the ID PROM.

In the SBus' earliest days, the ID PROM was not yet required. Instead, a 4-byte ID "register" provided a unique token (or "magic

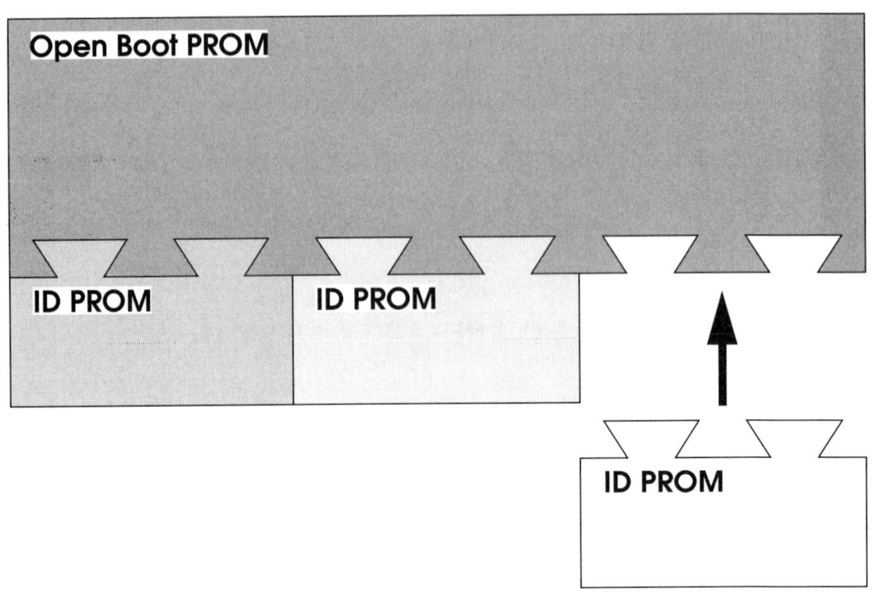

FIGURE 4.1. *The ID PROMs can Extend the Open Boot PROM.*

number"). This was used to index a table within the system's own firmware. This table contained the same information that the ID PROM does now. The ID Register scheme was abandoned because it was not practical to change the system's PROMs whenever a new device was added, nor was it practical to contain all possible attributes for all possible add-on cards (many of which might have been designed only after the host was already in the customer's hands).

Historical vestiges of this ID Register remain today. Some early SBus interface chips, such as LSI Logic's L64853(A), still contain such an ID Register. All SBus add-on cards must contain an ID PROM, however, and the ID Register should not be used.

4.1.3 Key Elements

The key elements of the Open Boot PROM are shown in Figure 4.2. This architecture contains three major elements. The first of these is the system-*dependent* core. The second is the Forth FCode interpreter, that provides the "language" that allows system-*independent* communication between the Open Boot PROM and the ID PROMs. The third element is a "toolkit" of utilities that provides common functions and allows a user to get in and interactively "tinker" with the machine.

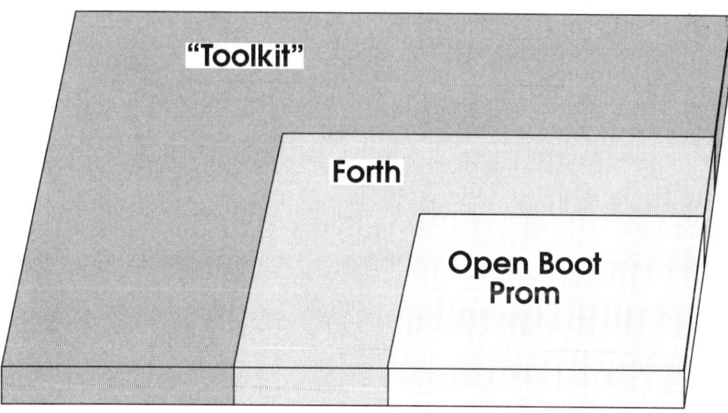

FIGURE 4.2. *Key Open Boot PROM Elements.*

System Dependent and Power-On Initialization

At the lowest level is system-dependent code that is responsible for initializing the host processor, testing memory, and performing other related functions that are common at power-up or reset. This code is invisible to the SBus interface, and can be implemented in any code or form desirable.

Forth (FCode) Interpreter

Contained within the system's firmware is a Forth interpreter. This interpreter is the mechanism used to maintain architectural independence. Forth is an interpreted language that does not make any assumptions about the underlying processor architecture. It is standard within the industry, compact, and efficient.

A card's firmware extensions are stored in a byte-tokenized version of Forth called *FCode*. This FCode is de-tokenized and executed by the Open Boot PROM at power-up or reset. This code is not used after the operating system and the card's driver have been loaded. It is used only to perform whatever low-level configuration and diagnostics are necessary, and to provide *attributes* to the OS that enable it to find, load, and configure the appropriate driver. In this way, the advantages and simplicity that an interpretive language provide can be had without the performance penalties that might otherwise result.

Toolkit

The *Toolkit* is also a part of the Open Boot Architecture. This is a built-in library of Forth functions that support the host's hardware, including the SBus interface. These functions allow registers and memory to be examined and modified, virtual memory to be mapped, and so on. The toolkit can be used as a console or monitor program that allows a user to enter commands directly and immediately view the results. Many of the toolkit functions can also be used programmatically by the FCode contained within an ID PROM, although some can only be used interactively (and some can only be used programmatically: for more information, see the *Open Boot PROM Toolkit User's Guide*, published by Sun Microsystems).

Interestingly, either the user or the ID PROM can also add "tools" of their own to the toolkit. New Forth "words" can be easily defined and added to the toolkit for all to use. This is the primary mechanism that allows new types of boot devices to be added to a host. Simple structures are defined to write a pixel, get or display a character, store or retrieve blocks of data, and the like. A new device can use its FCode to fill out these routines in its own device specific way. Then the host and other devices can easily use the new device at boot time, even if they have never before encountered it.

4.2 FCode and Toolkit

4.2.1 SBus Slot Probing

When an SBus based host powers up, one early action it takes is to "probe" the SBus slots. The ID PROMs on each card are accessed in turn, and the system's firmware builds data structures using the information it finds. These structures include information about the devices' names, interrupts used, and other attributes.

For historical and compatibility reasons, the ID PROM is queried with either 8-bit accesses, 32 bit accesses, or even a combination of the two (see the discussion on the now obsolete "ID Register" at the beginning of this chapter... this register responded only to 32-bit accesses). Any combination of access sizes is possible and the slave should be designed accordingly. One common scenario, though, is that the first access will be a 32-bit access. If this receives a byte acknowledgment (indicating that the operation is possible, but only with bus-sizing) then all subsequent accesses will be 8-bit accesses. Otherwise, 32-bit accesses will continue.

There is no minimum time from the de-assertion of RESET* to the first "probe." Cards which require additional time for on-board initialization must be aware of this.

The first byte of the ID PROM is expected to be the (hex) value 'FD'. In some very special cases, 'F0', 'F1', 'F2', or 'F3' may be found in the ID PROM's first location instead of 'FD'. These cases will be rare, and are used only when some unusual behavior is needed to read the ID PROM (if it was implemented with a serial EPROM, for example). As of this writing there is a working group (designated P1275) within the IEEE that is working on a standardized Open Boot Specification. Please refer to this for more information.

The system is prepared for a time-out when probing an SBus slot. If an error acknowledgment terminates the first probe of any slot then the machine will deduce that the slot is empty and then ignore it. A time-out is not the only mechanism that can generate an error acknowledgment, however. The slave may generate an error acknowledgment on this first access for any of a number of reasons. If it does so, it will be ignored. For this reason it is important that slaves be prepared for ID PROM transfers initiated with either 8-bit or 32-bit **SIZ** codes. If the slave properly accepts only one, and gives an error acknowledgment to the other, then it risks not being recognized in some machines.

If the system's first probe does not receive an error acknowledgment (resulting from either a time-out or some other problem), then it assumes that a device is present. It then begins to build a device tree entry, and if the data received matches the 'FD' value expected, then this entry is filled out with attributes contained in the ID PROM. If the data received does not match "FD", though, it cannot complete the device tree entry. The system's behavior in this case is less predictable.

All device tree entries at this level (and hence all ID PROMs) must contain a "name" attribute. If not, then SunOS will fail during the boot process. This will be one possible consequence of an incorrect (or missing) ID PROM. If this happens, the Open Boot Toolkit can be used to debug the interface to the ID PROM.

Another possibility is that the firmware will assume that a hardware fault of some sort has occurred. Remember that the machine is designed more for the end-user than for the developer.

If such a situation arose "in the field," then assuming that there is a hardware fault is the correct thing to do. It would be improper to simply ignore the slot and proceed. The error might not be just in the card itself. It might be in the SBus, or in the SBus controller, or some related element of the host. In such a case the host might not be able to proceed at all, and there is no guarantee that even an error message could be written out because the error might be in the frame buffer, or prevent access to it. The number of variations possible are enormous, and the ideal responses in each case vary considerably. The scope of this problem is far beyond what is practical to build into the system's firmware. As a result, the firmware may simply trap on the error, and may or may not try to write an error message to either the CRT or through one of the serial ports. In this event, the best recourse is to arrange for this slot to be skipped altogether during probing. This is done by setting the **sbus-probe-list** parameter as shown in Figure 4.3. In this particular example, the **sbus-probe-list** parameter is configured to probe slots 0, 3, and then 1, in that order. Slot 2 will not be probed at all.

```
ok  setenv sbus-probe-list 031          ...don't probe slot 2
ok
```

FIGURE 4.3. *How to Modify the "sbus-probe-list" NVRAM Parameter.*

4.2.2 Typical FCode Programs

An example of the minimum FCode program that must be contained in every ID PROM is shown in Figure 4.4. The "name" attribute, i.e. "MYCO,myproduct" is the string that is used within your driver's **identify** routine, enabling it to find and access the card. The "name" attribute is required. All ID PROMs must contain it, and it completely replaces the driver attribute which was (but is no longer) an alternative on early hosts.

The name attribute has two parts. The first (the "MYCO" string in this example) is a string whose purpose is to identify the vendor that produced the product. Any string of characters likely to be unique to your company will suffice. If your company is listed on any stock exchange, the stock symbol abbreviation is a good choice. The second part of the name attribute (the "myproduct" in this example) is the name of this product. It should be unique within your organization. There is no limit to the number of characters in either part of the "name" attribute.

```
fcode-version1
" MYCO,myproduct" name
end0
```

FIGURE 4.4. *An Example of a Minimum FCode Program.*

Notice that there is a space between the " character and the start of the text string **MYCO**. That space is a significant part of the FCode syntax, and must be at the start of any string parameter. This leading space is not actually stored as part of the string though—an important thing to realize later when it is time to attach the driver. Notice also that there is *not* a space after the comma in the name or before the closing quote. Inadvertent spaces here cause difficulties when trying to access the device using the **cd** command (discussed in more detail later in this chapter), so it is very important not to have any spaces other than the first, required one.

A more typical FCode program is shown in Figure 4.5. This program contains three additional features. Comments are incorporated. These are indicated by the \ character (followed by a space, which is significant and must be there).

The other two additions result in additional code. The first of these is the **reg** statement, which defines the card's register (addressable) spaces. Note that only one **reg** statement is significant. If more than one register space exists there should not be a separate **reg** definition for each. Multiple register spaces are handled by a more intricate variation of the **reg** syntax. See *Writing FCode Programs for SBus Cards*.

The last addition in this typical FCode example is the **intr** statement. In this case it informs the host that this card uses SBus interrupt level 4. Legal values for this parameter are 1 through 7, which correspond to the 7 SBus interrupt levels. The other parameter is a vector to associate with the interrupt. In most cases the null (0) vector is appropriate, as shown. As with the **reg** attribute, cards with multiple interrupts should not use multiple **intr** statements. A single, more involved variant of **intr** syntax should be used. See *Writing FCode Programs for SBus Cards*.

```
fcode-version1
" MYCO,myproduct" name
hex
\ The following line defines my register space. Only 1 "reg"
\ statement is allowed in an FCode program. If more than 1
\ register space exists then they must be defined simultan-
\ eously using a single "reg" statement:
my-address 100 +  my-space  100  reg

\ the following line states that I use SBus interrupt level 4.
\ the "0" parameter is a null vector:
4 0 intr
end0
```

FIGURE 4.5. *An Example of a Typical FCode Program.*

While the **reg** and **intr** statements contained in this last example are not absolutely required, certain operating systems (such as System V Release 4, or SVR4) may need them to properly integrate the device and its driver. It is wise to include them if possible.

These illustrations are only examples, meant to show that in many cases the level of FCode sophistication is very small. The actual FCode found in any given application may vary considerably, depending on the level of diagnostics or other functions built in to the ID PROM. Cards which can be used as boot devices especially will often require more sophisticated firmware. For more information the author strongly recommends *Writing FCode Programs for SBus Cards*, published by Sun Microsystems.

4.2.3 Boot Device Support

Boot PROMs with revisions 1.X do not support boot devices except frame buffers. Revisions 2.X do, first starting with the SPARCstation 2 class machines.

Boot devices require primitive "drivers" to be written for them in FCode. These "drivers" are actually a collection of custom Forth routines, though with pre-defined names and functions which the Open Boot PROM expects and understands. In this way the OBP can draw a pixel, blank the screen, or move the disk heads, for example, without having to know specifically how the device works. These routines provide only the most basic functions for the device. Several basic classes are recognized, including display, serial, disk, tape, network, block, byte, hierarchical, and so on.

As FCode is an interpreted language, performance of any device while using these low-level routines will be limited. These routines are only used during booting, however, until the "real" device drivers can be attached. After this point the FCode on the SBus card lies dormant, and does not affect device or system performance.

4.3 Survival Forth

This section is meant to teach just a few quick concepts and commands which will help test the basic functionality of an SBus card. The Open Boot PROM toolkit can be used to map virtual memory, examine and modify memory, and write or read devices on the card. Simple "scope loops" can be constructed, too, which facilitate using an oscilloscope to debug the card's logic.

The Open Boot PROM toolkit provides more options and commands than are contained here, and more detailed documentation is available from a variety of sources (see the Bibliography at the end of this book). What is contained here, though, is enough for anyone without Forth experience to "get their feet wet" (and "their hands dirty!") quickly and easily.

This information is not presented as a list of commands, parameters, and results. It is presented 'by example,' using traces of actual toolkit sessions (with a few interspersed comments). The reader is encouraged to get hands-on experience by repeating these examples personally. Branch off from them, too, whenever and wherever curiosity leads. "Playing" and experimenting with the toolkit might well be the best way to learn how best to use it.

4.3.1 Basic Concepts

One fundamental concept which must be understood about Forth (and hence the OBP toolkit) is that it is stack-based. Operands are pushed onto the stack, operations are performed on them, and results are then pulled from the stack. If you've ever used a calculator with reverse-Polish notation built into it, you've already been exposed to this concept.

A developer need only understand a few simple operations to manipulate the stack. Examples of some of the most basic stack manipulation operations are shown in Figure 4.6. To put a number

```
ok  clear                           ...clears the stack
ok  .s                              ...non-destructively displays the stack
Empty
ok  hex                             ...hexadecimal I/O mode (the default)
ok  53                              ...put (hex) value 53 on the stack
ok  .s
53
ok  d# 99                           ...this number is decimal, despite hex mode
ok  .s
53 63                               ...Top of Stack (TOS) is always right-most item
ok  decimal                         ...decimal I/O mode
ok  11 h# 17                        ...put 11 and (hex) 17 on the stack
ok  .s
83 99 11 23                         ...displayed in decimal this time
ok  drop                            ... removes Top of Stack (TOS)
ok  .s
83 99 11
ok  .                               ... removes, displays Top of Stack (TOS)
11
ok  113 .s                          ...more than 1 command per line is ok
83 99 113
ok  .s .s
83 99 113 83 99 113                 ...both displays are on the same line
ok  .s cr .s                        ...'cr' puts a blank line between displays
83 99 113
83 99 113
ok  dup                             ...duplicates TOS
ok  .s
83 99 113 113
ok  drop .s
83 99 113
ok  swap                            ...swaps the top two elements
ok  .s
83 113 99
```

FIGURE 4.6. *Basic Stack Manipulation Commands.*

on the stack, simply type it at the prompt. Numbers can be entered in either hexadecimal or decimal, depending on the default mode. This default mode can be changed whenever desired, or it can be temporarily overridden. Multiple numbers (and/or commands) can be entered at any prompt.

As with any other programming language, constants and variables can be defined. Some examples of how this can be done are shown in Figure 4.7. Once defined, constants or variables are used in just the same way as the numbers they represent. In any of the following commands that require parameters, the names of constants or variables can be substituted for the numbers that would otherwise be used.

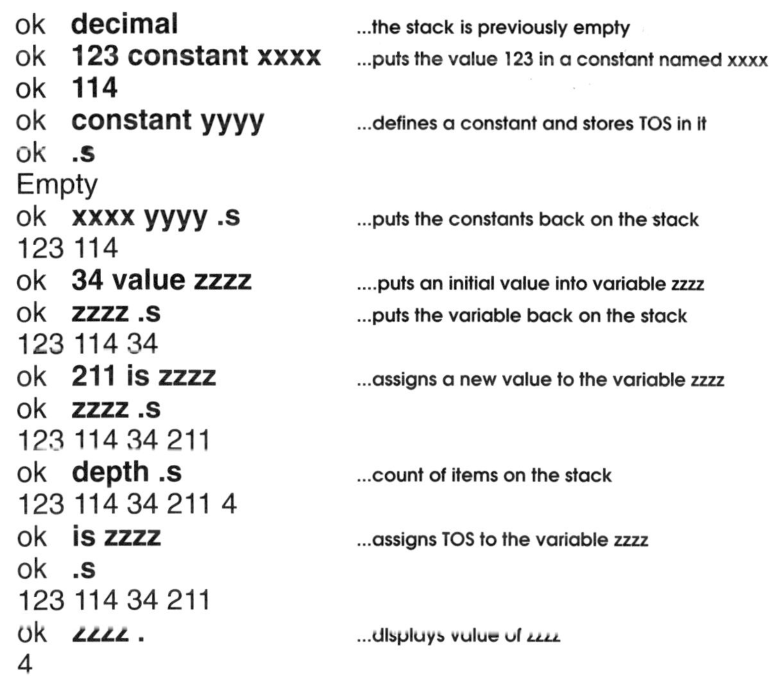

FIGURE 4.7. *Setting Constants and Variables.*

Note that Forth uses the keyword **value** to define variables, and many Forth programmers use that term to refer to variables. The author finds this confusing, however, and has chosen to call variables just that: variables. Attempt to read the following paragraphs while substituting the word "value" for "variable", and the reason will become clear.

Constants and variables differ in just the way one might expect: constants are intended to have values assigned to them only once, but the value of a variable can be changed as often as necessary. The reason a distinction is necessary relates to how the constant or variable is stored in memory. A constant can be stored in read-only memory (or in a page of memory that is mapped as read-only). Also, the actual value of the constant can be substituted for it when a word is being defined or tokenized, saving space and execution time. Variables can only be stored in random-access memory, and must be looked-up each time they are accessed, not just when defined or tokenized.

The value of a variable is changed using the **is** keyword, also shown in Figure 4.7.

Arithmetic is done on the stack. Usually, one or more values are removed from the stack, the desired operation is performed, and the result is left in their place. The most basic arithmetic operations and their results are shown by example in Figure 4.8.

4.3.2 Getting Started

If you are testing a new card design, or one with new FCode, probing might fail. In this case the machine will not boot, and depending on the failure mode it may not even be possible to use the toolkit to poke around and look at the card's on-board ID PROMs.

To avoid this problem it is a good idea to remove the slot that the card is plugged into from the sbus-probe-list in the machine's NVRAM. This is done using toolkit's **setenv** command, as shown in Figure 4.3. During any subsequent attempt to re-boot the machine the slot will be ignored instead of probed, and booting (or accessing the toolkit) should proceed normally.

Also, before FCode commands may be found and executed by name from the Toolkit, it is necessary to first set the NVRAM flag **fcode-debug?** true, and then reset the system, as shown in Figure 4.9

Use the **printenv** command to examine the state of any environment variable. If no argument is given then all environment variables will be returned. If the name of a variable is given as an argument, then only that variable will be returned. In either case, the display will include the variable name, its current value, and its default value.

```
ok   decimal 83 113 99 .s   ...the stack is previously empty
83 113 99
ok   -                      ...subtracts TOS from next element
ok   .s
83 14
ok   +                      ...add top elements
ok   .s
97
ok   12 * .s                ...put 12 on stack, multiply, display
1164
ok   7 / .s                 ...put 7 on stack, divide, display
166
ok   constant xxyy          ...defines a constant and stores TOS in it
ok   .s
Empty
ok   xxyy 10 + .            ...put constant, add 10, remove and display
176
```

FIGURE 4.8. *Basic Arithmetic Commands.*

```
ok   setenv fcode-debug?  true
ok   reset
```

FIGURE 4.9. *Setting the 'fcode-debug?' Environment Parameter.*

4.3.3 Mapping Virtual Memory

The SBus uses virtual addresses, so before you can access a device
in any way you must "map" it. This provides the memory management unit in the machine with a link between the virtual addresses
that you use and the physical addresses that the hardware (in particular the SBus Controller) must ultimately deal with.

Concepts like virtual memory, memory mapping, virtual-to-physical translation, etc., sound very complicated. Fortunately the OBP toolkit provides resources that greatly simplify virtual memory management.

The basic steps required to map SBus devices are shown in Figure 4.10. The first four lines are optional. These define constants which contain the physical address offsets of the various SBus slots. Note that this includes a definition for 'slot 0', which is not actually a physical slot, but is pseudo-slot that contains the motherboard's built-in devices (such as the ethernet and SCSI controllers). For readability purposes there are '.' characters embedded in long number sequences (such as '200.0000'). These are ignored when the number is interpreted, and can be placed anywhere. Convention usually places them every fourth digit, counting from right to left.

```
ok   000.0000 constant on-board        ... 'slot 0' devices
ok   200.0000 constant slot1           ...slot 1 address space
ok   400.0000 constant slot2           ...slot 2 address space
ok   600.0000 constant slot3           ...slot 3 address space
ok   slot1 1000 map-sbus        ...map a 1000 (hex) space into slot 1
ok   .s
ffee6480              ...the virtual address is left on the stack
ok   value myspace        ...it's useful to define a variable or constant
ok   myspace 120 +        ...slot 1, address 120 (hex)is left on the stack
```

FIGURE 4.10. *Mapping Virtual Memory.*

The values given in the constant definitions are not actual physical addresses. They are offsets into the system's SBus space, whose base address may vary from one machine to another. The offsets themselves may be system dependent, too, but this is far less likely; they contain a level of indirection which helps insulate the user from some system dependencies.

The actual mapping is done in the fifth line of Figure 4.10. The general form of this command is **<offset> <size> map-sbus**. The offset parameter, as explained, is the offset into the host's SBus address space. The size parameter is the size of the region that is to be mapped. Both of these parameters are placed on the stack before the **map-sbus** function is invoked. The result is a virtual address left on the stack in place of the two parameters. This virtual address may then be used to access the region just mapped. This can be stored in a constant or a variable. The use of a variable is often the better choice, because this allows the memory to be unmapped and re-mapped easily and as often as necessary.

As easy as it is to map memory in this way, it still requires a knowledge of which slot the device is plugged into, which is then used to determine a physical address offset which may vary from one machine to the next. This sounds counter-productive from an auto-configuration point of view, and it is! Fortunately, though, the OBP provides the constant **my-address**. This constant is pre-defined to contain the address offset for the card's slot, but only when used by FCode in the card's ID PROM. It does not give useful results when used from the toolkit's higher, interactive level. This is because it is obvious exactly whose address offset is needed when running out of the ID PROM, but from the toolkit the OBP has no way of knowing which slot or address you're after.

An example of how memory can be mapped from a device's ID PROM in a slot independent fashion is shown in Figure 4.11.

4.3.4 Address Limits During Booting

The total amount of memory that can be mapped from within the boot PROM and toolkit will often be limited. This limit is not the result of an SBus restriction, per se, but is the result of limited mapping resources in the system as a whole. The actual limits will vary from one system to the next. Typically, however, the Boot PROM will only be allowed to map something like one megabyte of memory. Much more than this would begin to degrade overall system performance.

Whatever the actual limit, this resource must be further divided up among the FCode diagnostic and initialization routines contained within the ID PROMs associated with each device. Any device that wishes to map a large amount of memory, to perform a memory test or other diagnostic, should do so by successively mapping and then un-mapping smaller "slices" of the overall space.

Fortunately this limit only applies during boot time, and even then it is still possible to declare much larger "reg" address spaces, which will be passed up to the device driver when it is loaded. Once the operating system is available and the device's driver is loaded, it can map as much memory as is necessary from the system's entire capacity.

```
FCode-version1
" SUNW,SBusProto" xdrstring " name" attribute
hex


\ These constants define which address space is to be
\ accessed
 0000000 constant prom-space
 0001000 constant prom-size
 0800000 constant rambuffer-space
 0008000 constant rambuffer-size

\ These constants define the base physical address of my
\ ID prom and ram buffer. The 'my-address' constant is pre-
\ defined within the OBP to provide me my address offset
\ into this machine's SBus space. I don't need to know what
\ slot I'm in!!
 my-address prom-space + value idprom-pa
 my-address rambuffer-space + value rambuffer-pa

\ This is where the two regions of memory I need (one for
\ the ID prom and the other for the ram buffer) are mapped
\ in. The resulting virtual addresses are stored in constants
 idprom-pa prom-size map-sbus
 constant prom-va
 rambuffer-pa rambuffer-size map-sbus
 constant rambuffer-va
```

FIGURE 4.11. *Mapping Memory From Within an ID PROM.*

4.3.5 "Peeking" and "Poking"

One of the major advantages that the toolkit provides is the ability
to examine and modify registers or memory on an SBus device.
This is often the first stage in debugging a board. Some of the commands and options available for doing this are summarized by
example in Figure 4.12.

```
ok   200.0000 4000 map-sbus    ...map slot 1
ok   constant mycard              ...set a constant (take value off stack)
ok   mycard .s                    ...put it back on stack, and display it
ffee5000
ok   c@                           ...do an 8-bit read, leaving result on stack
ok   .                            ...take result off and display it
fd
ok   a5 mycard                    ...put data 'a5' and the address on the stack
ok   c!                           ...do an 8-bit wide write
Data Access Exception            ...time-out, error ack, or map error occurred!
ok   mycard 1000 +                ...oops! Let's try an address past the PROM
ok   constant ramaddr             ...and define a constant
ok   a5                           ...put data 'a5' on the stack
ok   ramaddr c!                   ...put the address back and do the 8-bit write
ok   ramaddr c@ .                 ...look's good, let's test it
a5                               ...it worked!
ok   a5c3 ramaddr w!              ...16 bit write with data 'a5c3'
ok   ramaddr w@ .                 ...16 bit read
a5c3
ok   c3a53c5a ramaddr l!          ...32 bit write
ok   ramaddr l@ .                 ...32 bit read
c3a53c5a
```

FIGURE 4.12. *Basic Read and Write Operations of Various Sizes.*

The "c" in the 8-bit commands is short for 'character'. The "w" in
the 16-bit commands is short for 'word'. This seems odd because a
"word" in most SBus machines is 32-bits wide. Forth's basic syntax
dates from a time, however, when 8-bit machines were the norm
and 16-bit addresses or operands were indeed often called 'words'.
The "l" in the 32-bit commands is short for 'long-word'.

4.3.6 'Scope' Loops

These are small tight loops of Forth code which repetitively access
the card in some known way. Even though this code is interpreted,
the repetition rate is usually high enough that good traces can be
obtained on standard oscilloscopes. This makes it significantly eas-

ier to trace the propagation of addresses, chip-enables, and data through the card.

There are a few points that should be made about the syntax shown in Figure 4.13. The **begin .... until** construct is one that will loop, executing all the commands between the two keywords, until the value on the top of the stack is non-zero (FCode considers non-zero values 'true' and zero values 'false'). In testing the value at the top of the stack it is removed and discarded. The **key?** command is a function that returns a non-zero value if a key has been pressed, and zero if one has not. Hence, **begin .... key? until** loops until a key is pressed. Also, **key?** puts a value on the stack, but the test implicit in the **until** keyword takes it right off again, so the stack is only affected by the commands inside the loop, not the loop itself.

```
ok   200.1000 1000 map-sbus constant myreg
ok   begin myreg c@ drop key? until     <loops until any key>
ok   begin a5 myreg c! key? until       <loops until any key>
```

FIGURE 4.13. *Simple Read and Write 'Scope' Loops.*

The **drop** command is necessary in the read loop because the fetch command (**c@**, **w@**, or **l@**) replaces the address on the stack with the data fetched. If something were not done to remove that data then the stack would quickly overflow and the machine might cease to behave in a useful manner (translation: you've got a 'crash-and-burn' coming). The '**dot**' (.) command could be used to remove the data and print it. This might be very useful as an indicator that the machine is still behaving properly. It might also be a useful way of displaying continuous readings from a changing data source. The drawback, though, is that it greatly slows down the loop's repetition rate.

How might such a scope loop be used? Suppose that the SBus card isn't working properly; it returns **ACK**s when read or written to, but the data that is returned on reads is not correct. To debug such a problem, a read loop could be constructed like that shown in Figure 4.13. While this loop is running, an oscilloscope could be triggered from any signal which you expect to be active throughout the cycle. A good example might be the chip select line of the memory, register, or other device that you are trying to access. If this select signal is not occurring as expected, that is a good clue that something is broken in the decoding logic: the scope could then be used to trace the fault back towards **AS***, **SEL***, and the physical address lines. If the select signal is occurring, it will help provide a stable scope display of the cycle's anatomy. Another scope probe can then be used to make sure that data buffers are enabled and pointing the right way. It can be used to check that read or write strobes are in the proper state. It can even be used to make sure that the address and data at the device are correct. Quite a lot of information can be had in just a short time, and with simple tools. Also, the scope loop can be easily modified to change addresses or data patterns, etc.

Repetition rates of about 2 KHz are typical for scope loops such as this (depending on the type of host, its clock rate, and other factors). This will be fine for many applications where the scope sweep rate is moderate. In some situations very high scope sweep rates are necessary, though. This would be true if, for example, a signal's edge or a narrow pulse were being examined. In cases such as this, a higher repetition rate may be necessary to guarantee that the scope display's phosphors are activated strongly enough to provide a suitably bright display.

Increasing the repetition rate is straightforward. The strategy is to reduce the overhead associated with testing for a key and then looping on the resulting flag. One method is to imbed a simpler, more efficient loop in the scope loop. An example of how this can be done using a 'do' loop is shown in Figure 4.14. A **do** loop is a simple construct that repeats the series of instructions contained between the **do** and the **loop** keywords. The two parameters that precede the loop are the final and the initial value of the loop's index, respectively. (For advanced students, the loop index may be accessed within the loop using either the **i** or **j** variable, depending on the nesting level.)

```
ok  begin 20 0 do a5 myreg c! loop key? until
```

FIGURE 4.14. *'Scope' Loop with Nested 'Do' Loop to Improve Repetition Rate.*

This improves the repetition rate because most of the inner loop's overhead is only an increment or decrement operation, which is much quicker than the I/O operation that must be performed in the outer loop to check for a key pressed. In this particular case the write operation will be performed 20 times for each outer loop. The resulting repetition rates can be 20 KHz or more.

Another method involves copying the write or read operation within the loop. An example of how this might be done is contained within Figure 4.16. This slows down the overall loop only slightly, but now multiple accesses will occur at rates that can approach 40 KHz within each loop. The scope's phosphors will average out the burstiness of the resulting accesses, and provide a much better display as a result.

4.3.7 Define Your Own "Words"

Toolkit commands such as **map-sbus** are commonly called "words" (borrowing Forth's terminology). These commands are really functions or procedures much like the subroutines that can be found in almost any programming environment. They are usually built-up out of simpler functions, and can be used to build ever more complex commands.

An example of how a custom word can be defined is shown in Figure 4.15.

```
ok  : test-read              ...the ':' starts the definition, then we name it
   ] begin myreg c@ drop key? until
   ] ;                        ...the ';' ends the definition
ok   see test-read           ...'see' is used to disassemble any word
 : test-read
   begin
     myreg c@ drop key?
   until
 ;

ok  : test-write begin a5 myreg c! key? until ; ...one line is ok
```

FIGURE 4.15. *Defining A 'Custom' Toolkit Word.*

A word-definition is started with the ':' character, which is actually a Forth word in its own right. Follow that with the name of the new word, and then with the definition of the new word. Finally, terminate input of the new word by typing the ';' character (also a Forth word in its own right).

The new word definition may be typed all on one line, or it may cover several lines. While in definition mode (initiated by typing ':'), every carriage return will result in a ']' prompt. This will continue until ';' (ending definition mode) is typed. Definitions may be as long as necessary (within the limits of the toolkit's available "dictionary" space. Word editing capabilities are limited, however, and so it is advisable to build up complex definitions out of smaller, simpler ones. This eases modifications and usually reduces the overall amount of typing necessary. The **see** command can be used to disassemble the contents of any word. An example of its usage is also shown in Figure 4.15.

It may be important to understand some of the mechanisms Forth uses to store words defined in this manner. All words are kept in a "dictionary." This dictionary is not sorted alphabetically, though, but chronologically: words are stored in the order they are defined. It is possible for a word to be defined multiple times intentionally or accidentally (see section 4.3.9). If this happens, then only the most recent definition will be invoked whenever the word is used.

As recommended above, any complex word definition may be built using other, simpler words. It is much easier to build and understand functions in this hierarchical way. Whenever this is done, the new definition being formed captures and stores snapshots of all words invoked. This is an on-line compilation, of sorts, and it greatly improves the performance levels that can be achieved. There are interesting side effects, though. If a word is re-defined (becoming multiply defined, as discussed above), then any *future* use of the word will use its new definition, as expected. But any other word that has been *previously* defined using this word will still contain a snapshot of the *original* definition, and behave accordingly!

4.3.8 Put These Commands in PROM

It would be cumbersome, of course, to redefine custom toolkit "words" every time the system is brought up. It can even be cumbersome to retype the simple mapping and scope-loop commands if you must power up and down frequently to change chips, add wires, inspect rework, or the like.

There is a simple remedy for this, once the board is debugged enough that the ID PROM can be read. Put these definitions in the ID PROM (somewhere after the minimum FCode program described earlier in this chapter), and they will automatically be loaded at probe time! In this way, it is as if the debugging commands are now an integral part of the toolkit, available whenever needed. An example of how this might be done is shown in Figure 4.16. A scope loop is included named **test-my-idprom**. This loop performs continuous reads to the ID PROM space, which can be used to help debug the logic involved in such an access (the fetch is repeated 3 times per scope loop to insure a high repetition rate, as discussed previously). Note that this particular example might not actually be as useful as some, however. For this code to be loaded from the ID PROM and made available for use within the toolkit, the ID PROM must be present, functioning, responsive to both 8 and 32 bit accesses, and contain the proper expected value in the first byte location. If all of this works then there is little need for a

```
FCode-version1
" SUNW,SBusProto" name
hex


\ The constant my-idprom is used for tests of the prom
\ space
my-address 1000 map-sbus constant my-idprom


\ This is a scope loop that can be used to help debug
\ the ID prom interface
: test-my-idprom
        begin
                my-idprom c@ drop
                my-idprom c@ drop
                my-idprom c@ drop
                key?
        until
;
end0
```

FIGURE 4.16. *Sample FCode with Built-in Debugging "Hooks".*

scope loop to test the ID PROM space. If any of this doesn't work, though, then such a scope loop might be helpful but will have to be typed in by hand or downloaded from a file.

Once the ID PROM is functioning, though, it can serve as a very useful tool when debugging the rest of the board. It can be used to better and faster initialize the debugging environment, saving the engineer or technician many keystrokes.

4.3.9 Avoiding Name Conflicts When Defining Custom Words

When defining a custom word it is important to pick a name for it that is unique. Otherwise, the new definition may supersede an already existing one, preventing the original from being accessed or used (this is a simplification; for a more complete description of the process involved, see section 4.3.7). Fortunately a warning message can be expected if a new name is not unique.

Open Boot revisions above 2.0 include features which are designed to reduce the chances of name conflicts. Names are qualified with their relative location on the device tree, so that they must only be unique on a device-by-device basis. This device tree can be navigated in much the same way that a UNIX file system is. In fact many of the same commands are used! The **cd** command is used to change the current node. The command '**cd ..**' can be used to move to the parent of the current node, as might be expected. The **pwd** command is used to show the name of the current node, and the **ls** command is used to show daughter nodes. Daughter nodes are named using the **name** attribute associated with the device. The slot number is also used (slot 0 refers to a device on the motherboard), so that names remain unique even if more than one of any type of device is installed in the same host. Examples of how several of these commands might be used are shown in Figure 4.17. The **show-devs** command lists all existing devices, and can be used from anywhere within the device tree.

In general, you must **cd** to a device before you can list the words associated with a device (using the **words** command), and before you can invoke them. Note, however, that in some cases simply using **cd** to move to the proper location may not be enough to ensure proper functionality. It may first be necessary to select the associated device with the **select-dev** command. This command performs 3 vital functions. First, it sets up the toolkit environment so that function calls such as **my-address** return the proper values for this device. Second, a device specific routine named "**open**" is

```
ok  cd /                          ...go to the parent (root) node
ok  ls                            ...lists all "children" of root node
ffeb4c3c options
ffeb2aa8 fd@1, f7200000
.

.
ffea5e2c sbus@1, f8000000
.

.
ffea4a74 packages
ok  cd sbus                       ...go to the sbus node
ok  ls                            ...list all "children" of sbus node
ffeba91c bwtwo@3,0
ffeb9034 le@0, c00000
ffeb5b20 esp@0, 800000
ffeb5adc dma@0, 400000
ok  cd @0,400000                  ...devices can be accessed by slot, offset
ok  cd /sbus/le                   ...devices can be addressed by name
ok  .attributes                   ...list this device's attributes
mac-address                       08 00 20 0e 35 5a
device_type                       network
.

.
alias                             le
name                              le
ok  pwd                           ...show our current location
/sbus@1, f8000000/le@0, c00000
ok  words                         ...list the associated words
reset    seek      open    close    load
selftest write      read    le-package
le-selftestwatch-netset-vectorsle-pollle-xmit
init-obp-tftp
More [<space>, <cr>, q] ?
```

FIGURE 4.17. *Navigating Through the OBP2 Device Tree.*

executed, performing any necessary initialization. Finally, a **cd**
operation is performed to this device, eliminating the need for any

further command qualification. An example of how to use the select-dev command is shown in Figure 4.18.

---

ok " /sbus/my-device-name" select-dev

FIGURE 4.18. *An Example of How to Use the OBP2 Select-Dev Command.*

Acknowledgments

Special thanks to Mike Saari for all his help on this chapter.

References

Writing FCode Programs for SBus Cards, Sun Microsystems, 1990.

Pitfalls to Avoid

5

The purpose of this chapter is to summarize some of the key areas
where SBus developers may encounter difficulties in their efforts
to build SBus products. Much of the information here is distilled
directly from experience; (occasionally, very painful experience). By
providing the information here, it is hoped that new product developers can benefit from that experience, and produce better quality
products that reach the market faster.

5.1 Technology Issues

The following sections describe issues related to the technology
used on the SBus. In a way these concerns are not SBus specific,
because they are valid for any circuit or interface that uses this
technology.

5.1.1 CMOS Latch-Up and ESD Sensitivity

In the past, CMOS technology has had a (sometimes well deserved)
reputation for being sensitive and tricky to work with. Modern
CMOS has eliminated many of those problems. Still, CMOS is different than bipolar and other technology types in many ways, and
a few precautions are wise for those unfamiliar with it.

One element to consider is electro-static discharge (ESD).
Static charges exist everywhere, especially when the air is dry.
Everyone is familiar with the spark and shock that can sometimes
be felt after walking across a carpet and touching a doorknob, for
example. ESD damage can occur even when static levels are too
small to be felt. Discharges can blow out the metal used to connect
I/O pads on integrated circuits, or it can breach the transistors
connected to them. Sometimes, the damage is not immediately
fatal, but will cause the part to fail prematurely at a later date.

Subtle ESD damage has often been linked to excessive "infant mortality" rates.

All integrated circuits are susceptible to ESD damage, but the transistor structures and fine geometries used in CMOS make it especially sensitive. Most parts now include protection diodes which help alleviate the problem a great deal. These work by breaking down under the high voltages that occur during a discharge, and shunting the resulting current directly to the ground or power rails. These diodes are often quite robust, but even they can be damaged under extreme conditions. It is far better to avoid such discharges in the first place. The best way to do this is with careful handling and proper equipment. Use the anti-static bags and foam that parts and boards are packaged in. Also use anti-static benches, floor-mats, soldering irons with grounded tips, and so on. Grounded wrist straps or ankle straps are important, too (for safety's sake make sure these include a high-value series resistor, and that they are unplugged before running for the phone). Consider using room humidifiers or ionizers to reduce static buildup, and ban styrofoam cups, fuzzy acrylic sweaters, and other major sources of static electricity. The author once worked for a company that spent thousands of dollars on anti-static equipment, training, and procedures, without realizing that every evening the janitorial staff came in and rubbed-down the test stands with acrylic (highly static-prone) dust mops!

CMOS components are also sensitive to a condition known as *latch-up*. The substrate forms part of a large-scale p-n junction, which is a side-effect of the fabrication process. Normally this junction is reverse biased and does not effect the operation of the part. In extreme cases, however, it is possible for this junction to become momentarily forward-biased. If it does, the junction may behave much like an SCR and continue to sink current until all biases are removed. In essence, the part will "latch" itself into an unusual mode where current consumption becomes excessive and proper operation is highly unlikely. If allowed to persist, long term damage may result.

The conditions necessary to induce latch-up vary, but the most common cause is an input or output pin that is driven too far above VCC or below GROUND. This might occur in several ways. For example, if logic levels were applied to a part before power was, then latch-up might occur. Excessive overshoots and undershoots may also cause latch-up.

Again, modern CMOS is a lot less sensitive than earlier relatives. Input protection diodes help here, as well, by limiting an input's ability to stray outside of the power rails. Still, the best cure is an "ounce of prevention." Care should be taken in systems where power can be applied to different parts at different times. Also, overshoot and undershoot should be minimized wherever possible (this last is a good idea for many reasons, of course).

5.1.2 Holding Amplifier Metastability Problems

Holding amplifiers may be used in some hosts instead of high value pullup resistors. These are described in detail in section 3.6.3 on page 89. In brief, though, the primary purpose of holding amplifiers is to prevent signals from drifting and loitering near input thresholds. Otherwise excess power dissipation might result—or even, in rare cases, unreliable operation.

A holding amplifier is a weak latch that holds a signal in its last valid state, whatever that state may be. Holding amplifiers do not sample a signal relative to a clock or any other signal, so they do not have setup times per se. They do have hold times however. An input transition must gain a new logic level and then hold it for a minimum period of time before the holding amplifier can see or record it reliably. If the hold time isn't satisfied, or if the input does not adequately reach the appropriate threshold, then metastability may result. This might cause the very situation that holding amplifiers were meant to prevent!

This problem is likely to occur in one of several ways. If the line connected to a holding amplifier undergoes a very short glitch, for example. This is unlikely to cause problems, though, because as the driver once again settles to the correct level, the holding amplifier will quickly follow.

Another possibility is that the technology used may not be "well-behaved," in that the process of tri-stating the output might momentarily cause the output to drive a different or illegal logic level. This would be serious if it occurred, but is unlikely because most technologies and logic families are carefully designed to behave reliably and predictably.

A much more dangerous scenario occurs if the signal's driver starts to change its state, and then tri-states the signal somewhere near mid-swing. In this case it is highly likely that the holding amplifier will then become metastable. The conditions required to cause this to happen may seem like an improbable scenario, but it might actually occur relatively frequently.

For example, suppose part of your circuit looks like that shown in Figure 5.1. Here an external 74FCT245 bidirectional transceiver is used to buffer an SBus data line. A close look at the data sheets (from a variety of vendors) for this buffer reveals that the output disable time can exceed the data propagation delay. The same is true of many similar parts.

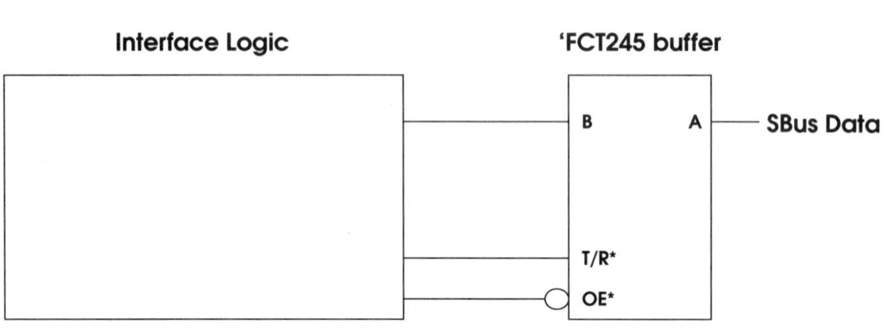

FIGURE 5.1. *Example of a circuit which may interact negatively with holding amplifiers.*

If the unspecified interface circuitry allows the data to change before the buffers output enable (OE*), or if they change at roughly the same time (as they often would in synchronous designs), then the buffer could start to drive the changed data onto the SBus before it tri-states. If the timing is unfortunate the buffer might not be able to complete the signal swing before it tri-states, and this is where the problem arises.

If you are doing any kind of SBus design, try to ensure that your output drivers do not ever partially drive a signal on the SBus. This is a good idea even if holding amplifiers are not used in the host, because it cuts down on unnecessary transitions and signal bouncing.

If you are building a host that uses holding amplifiers, though, do not assume that all drivers will be perfectly behaved. This potential problem was not discovered until there were already a substantial number of existing SBus designs on the market. These may or may not partially drive signals. Therefore, it is important to remember that some boards may cause metastability problems with holding amplifiers.

Fortunately, this is usually not a fatal situation should it occur, and it is probably sufficient for card and host designers to work to limit the exposure wherever possible. Holding amplifiers provide enough of a benefit in some applications that they are a good trade-off despite this problem.

5.2 Specification Issues

There are several instances where the SBus Specification is ambiguous or incomplete. There are some cases where it is somewhat out of touch with reality. There are some parts that, for whatever reason, are often ignored or stretched. This section is meant to provide some information and background on these "loose ends," to help an SBus developer maximize their chances of success.

This section should not be construed as permission to violate the specification in certain areas. Far from it. It is very important that designers make every effort to design within the Specification. This, in turn, will maximize compatibility and enhance the SBus overall; a goal that ultimately rewards all SBus developers.

The information contained in this section stems from the fact that like any human enterprise, the SBus Specification is not perfect. Further, SBus designers aren't perfect either, and are subject to making design mistakes and engineering trade-offs based on flawed judgment and information. Recognizing that, it is sometimes necessary to practice "defensive designing." By understanding the mistakes and trade-offs other designers have made, you are better able to avoid them yourselves and make your own design as inter-operable as possible.

5.2.1 Excessive Current Draw

Some SBus cards draw more current then the specification allows. Most often, this is due to simple mistakes. Rarely, a designer who feels pushed into a corner may choose to violate the limits by "just a little bit," or he may use statistical or typical values instead of worst-case.

Whatever the cause, violations of the +12 volt and -12 volt supplies are the most commonplace. For example, true Ethernet adaptors are likely "criminals," because they must supply 500 mA to the "mux-box" transceiver on the Ethernet coax. This is more than an order of magnitude greater than the 30 mA allocated to a single width card!

The power supplies in most hosts often have excess capacity. The host may not be fully configured, either, and some elements may not use their entire allotments, leaving more for those that use up theirs, and then some. It is not possible to determine exactly when or if a failure might occur, but the possible consequences of exceeding the machine's power budget is easier to pin down. If overloaded, a power supply may drop out of regulation or shut down altogether. Even if the power supply can handle the excess load, the host's cooling system may not be able to. It's even possible that traces would be blown, or ground shifts may be high enough to reduce your noise margins and cause faulty operation.

If you are an option-card designer, be careful not to exceed your current allotment. If you are a host designer, remember that not every designer is perfect and that excessive current may be used: give yourself some head-room.

5.2.2 Slaves might Drive Acknowledgments past the 255th Clock

The SBus Specification allows a slave to drive the **ACK*** signals as soon as the cycle after which **AS*** and **SEL*** are asserted. At this time they might either be driven with an acknowledge code that will end the cycle, or they might be driven with an "idle" code that serves as a placeholder until the slave is actually ready to complete the cycle.

In either case, the specification mandates that the slave does not drive these signals beyond the 255th clock in a transfer. This requirement exists so that the SBus controller may come in on the 256th clock and "time-out" the cycle with an error acknowledgment.

Unfortunately, most slaves do not contain logic to guarantee this because it is regarded as a lot of logic to throw at a problem that should happen infrequently (if at all) during normal operation. Further, even those slaves that do attempt to guarantee that this requirement is satisfied can't always be trusted because a failure severe enough to cause the time-out in the first place may hamper safeguards used to prevent this problem.

If you are building an SBus based host, you might consider using **ACK*** drivers potent enough to overdrive at least most of the **ACK*** drivers likely to be used on slaves. If you are building a slave, try to guarantee this requirement, and use weak drivers in any case (or put series terminations on the drivers, to increase their output impedance).

5.2.3 Leakage Current Is Often Fudged

PALs outputs driving the SBus directly often have off-state leakage currents of 40 µA, which exceeds the specified maximum of 30 µA. Earlier versions of the specification only allowed 10 µA leakage current and this problem was of great concern then. At that time the problem was researched, and it was discovered that some existing SBus designs had leakage currents as high as 130 µA in some cases! An analysis was done to determine the actual limits that could be tolerated. The 30 µA number now found in the specification is the result. It was understood that PALs would likely continue to be used, and that they would slightly violate the specification. Despite this, 30 µA was chosen because it covered the vast majority of logic families that seemed appropriate for the SBus, and it was desired to encourage people to seek other, lower-leakage alternatives whenever possible.

The bottom line is that you probably needn't sweat the extra 10 µA if you really DON'T have any alternatives. Try to make sure your typical number, at least, is less than 30 µA. When choosing your drivers, too, you might also want to consider whether they'll work if every load has fudged this factor and gone to 40 µA.

Don't push much beyond the extra 10 µA in any case.... This isn't a situation where there is a hard limit between a working design and one that fails. It's all to easy to get lulled into believing that "just a little more won't hurt." The further you go beyond 30 µA, though, the more likely it is that there are other alternatives you've ignored and the less sympathy you'll get if something breaks. In fact anything above 30 µA is a gamble, and one you should make sure you're willing to take. The penalties for losing this gamble could include a flaky design, configuration headaches, and ultimately a shoddy reputation.

5.2.4 Minimum Rise/Fall Times Are Often Ignored

The SBus Specification requires that signal rise and fall times be no less than five nanoseconds in most cases (only the clock is an exception; its minimum rise or fall time is 1 nanosecond). This was specified because the SBus is not terminated and must act as much like a lumped load as possible. In order to achieve this, instead of the transmission-line alternative, signal propagation delays must be kept to some fraction of the signal rise and fall times. This can partially be accomplished by keeping the signal runs very short, but it also requires reasonable edge rates.

This minimum rise/fall time specification is difficult to achieve, however, because there are only a very few slew-rate limited technologies that are available and appropriate. Further complicating the matter is that there is no minimum capacitance specification, and so modelling a worst-case situation cannot be done precisely.

Unfortunately, the only real solution is often to simply do your best to limit edge rates. Try to steer away from very high-power, high slew rate technologies. Series terminations can help, too, because they will combine with distributed capacitances to slow down the edge rates. For example, a 25 Ohm series resistor driving a 160 picofarad load results in just about a 4 or 5 ns rise and fall time!

5.2.5 Clock Skew and Capacitive Loading Violations

The SBus specification allows for up to 2.5 nanoseconds of skew. In most cases this shouldn't be difficult for a host to guarantee. It is even possible that future revisions of the specification will reduce this value to as little as 1.5 ns; the rationale is that the original value is too "sloppy," and that the time allocated is better "spent" elsewhere.

One very important factor in SBus clock skew is the variation of the clock load's capacitance. The more variation there is, the more clock skew results. The following formulas show the approximate relationship of varying capacitive loads to clock skew. The relationship varies depending on whether the clock distribution circuit behaves like a lumped load, or a transmission line. These formulas are approximate because they assume a 0 ns rise time at the driver, a perfect transmission line (if applicable), and a purely capacitive load:

$$\text{Lumped Load: } \Delta T = R * \Delta C * \ln(1 - (V_{th}/V_{oh})) \qquad \text{(EQ 1)}$$

$$\text{Transmission Line: } \Delta T = R * \Delta C * (V_{th}/V_{oh}) \qquad \text{(EQ 2)}$$

Where ΔT is the additional skew contributed by this factor, ΔC is variation in load capacitance, R is the source impedance (or source + line impedance in the transmission line case), V_{th} is the receiver's threshold voltage, and V_{oh} is the driver's (loaded) output swing. (If you are working with an expansion card, use the transmission line model, assume R=75 Ohms, V_{th}=1.5, and V_{oh}=2.4).

Rev. B.0 allows all inputs (including the clock) to have no more than 20 pF capacitance; there is no minimum. Future revisions of the Specification will probably set a minimum, though, and a likely value is 12 pF (this is a typical minimum load today). Plugging in the numbers, a variation between 12 and 20 pF at the load adds about 0.4 ns to the clock skew. That's a big factor when compared to the entire 2.5 ns (or 1.5 ns!) skew budget.

These formulas emphasize the need to avoid clock loading violations. The SBus clock is difficult to buffer, so it is not uncommon to find expansion cards that "cheat" a bit, connecting multiple loads to the clock pin on the bus connector. This results in load capacitance values that may be double or triple that allowed in the specification. If the load capacitance did double (from 20 pF to 40 pF), this would add almost one full nanosecond of additional delay to the clock on this board. That one nanosecond must be subtracted from the hold time of any signal that this board receives, and from the setup time of any signal that this board drives.

5.3 Protocol Issues

The following sections contain information about the SBus protocols. This includes some general guidelines, and then more specific discussions of issues related to bus sizing, burst operations, and atomic operations.

5.3.1 General Guidelines

This section contains general guidelines about how to use some of the SBus signals, and what they really mean.

Synchronously Sample SBus Signals

SBus signals which are synchronous should be sampled synchronously. This sounds obvious, but a brief look at what it means is useful because it will help to eliminate some common misperceptions and design mistakes.

Strictly speaking, a synchronous signal is one that is valid only on clock edges. Setup and hold times with respect to the clock edge are guaranteed, but no assumptions are made about the signal's state outside of this window. The signal may glitch, momentarily float, temporarily de-assert and then re-assert, or remain asserted longer than absolutely necessary.

These behaviors are not the result of careless design. They can be natural consequences of differential delays, state machine designs where outputs are combinatorial functions of state bits and the inputs, etc. Some or all of these actions would be serious problems in an asynchronous bus environment, though, and very careful design to eliminate such problems is often required in those environments. The desire to avoid such stringency and sensitivity is one reason the SBus is synchronous in the first place. Synchronous designs are easier and more robust than asynchronous designs when done properly.

One consequence of synchronous state machine design, though, is that synchronous signals should not be used asynchronously, or "raw." All such signals (or the results of any combinatorial function using them) must be clocked into some state bit on the next rising edge of the clock.

Not all SBus signals are synchronous. The interrupt request lines are completely asynchronous and level-sensitive. The **DATA**, **READ**, **SIZ**, and physical address lines may be considered glitch-free levels during the time in which they are valid (except during 64-bit Extended Mode transfers). **AS***, **SEL***, **BR***, **BG***, **LERR***, and the **ACK** signals are all strictly synchronous, though. **AS*** and **SEL*** are especially critical, as discussed in the next section.

Qualify SEL* with AS*, Synchronously

The **SEL*** signal is only valid when qualified by **AS***. If **AS*** is not asserted then no assumptions should be made about **SEL***; it might be asserted, de-asserted, in transition, or glitching. **SEL*** is known to glitch on certain machines; a natural result of the ripple-through logic which generates it.

As discussed in the previous section, too, **AS*** and **SEL*** are both synchronous signals. They are guaranteed only for the setup time and the hold time that surrounds each rising edge of the clock. No assumptions can be made about either of these signals outside of that window. The proper relationships and relevant assumptions concerning **CLOCK**, **AS***, and **SEL*** are shown in Figure 5.2. Note that **SEL*** is valid only when **AS*** is asserted, and only within either one setup time or one hold time of the rising edge of the clock. It is either invalid or irrelevant at all other times.

If **SEL*** is not first synchronously qualified with **AS***, then the slave may function erratically. For example, let's consider what would happen if the **SEL*** were used directly to enable the slave's data output buffers. This might seem reasonable at first, because **SEL*** is used to indicate to a slave that it is being accessed. This will not work, however. It might allow the slave to inadvertently drive and corrupt the data lines during some transfer (or translation cycle) in which it shouldn't be involved. Multiple slaves might conceivably be enabled at one time. **SEL*** glitches could also cause this slave's buffers to momentarily turn off during a transfer, which might make it impossible to transfer the data or meet the required setup and hold times.

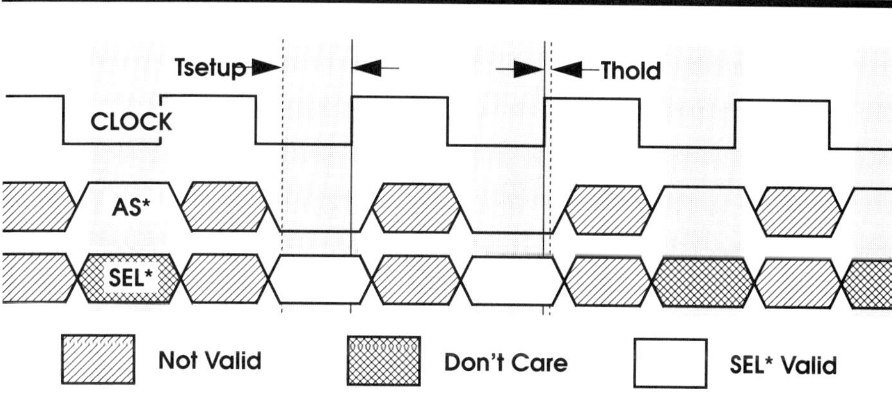

FIGURE 5.2. *The Proper Relationship Between CLOCK, AS* and SEL*.*

Always synchronously qualify **SEL*** with **AS***. Just two of many possible methods of doing this are shown in Figure 5.3. In each case, the signal **MY_SEL*** is asserted only when the slave is being accessed and is free to drive the data bus. It is also valid only when the **SIZ**, **READ**, and physical address lines are valid. The left-most of these circuits exhibits the minimum setup time in most cases. Unless this is an issue, though, the right-most circuit will usually be preferable for several reasons. The **MY SEL*** signal will be available earlier, for example, and will be less likely to produce glitches. Also, it requires less logic, and is less likely to suffer metastability if **SEL*** misbehaves between transfers.

AS* and SEL* Might Stay Asserted after the Cycle
AS* and **SEL*** may not be de-asserted immediately after a cycle. One or the other might stay asserted for several clock cycles after the transfer has ended. This is due to delays in the SBus Controller's state machines.

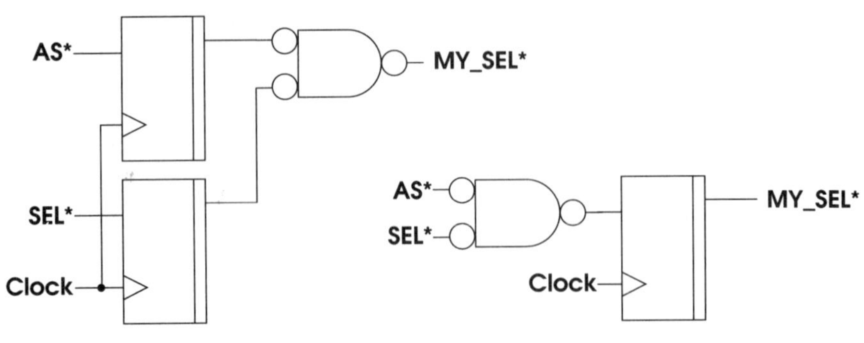

FIGURE 5.3. *Some Selection Circuits Which Properly Use **AS*** and **SEL***.*

Care must be taken that a slave's state machines do not mistake this situation for the start of a new transfer, rather than the end of the last one. For example, if the slave's state machines proceed directly to an "idle" or "ready" state once the transfer has completed, the slave could easily sample **AS*** and **SEL*** asserted on the next clock. Incorrectly assuming it was selected again, it would begin to perform a new transfer that isn't necessary and can only cause problems.

It is probably a good idea to design your slave's state machines so that they wait for **AS*** to be de-asserted (guaranteed for at least one clock between transfers) before re-entering its idle or ready state. This can be seen in Figure 5.10. There are two states in this transition diagram that proceed directly to the idle state, but only if **AS*** is not asserted.

AS* Might Not Occur

It is possible that an SBus transfer cycle may be started and completed without **AS*** ever being asserted on the bus. The SBus controller may end a cycle with an error acknowledgment if it detects an error (such as a problem with an address translation). It may not ever drive **AS***, however.

Even more likely, the targeted slave might be logically but not physically attached to the SBus. There are many ways this could occur. On-board devices or system memory are two examples of slaves that might have special "back-door" connections to the SBus controller, and so not need a direct attachment to the bus. Another possibility is that the targeted slave could be on another bus which is connected by some form of bus-bridge. This could be another parallel SBus, or it could be MBus, VME, Futurebus+, etc. In any such case it would not be necessary to assert **AS***.

The SBus Specification requires that masters rely on **BG***, not **AS***, for all of their sequencing (page 45 in revision B.0). SBus slaves must use **AS***, of course, but the only possible problem here is that the slave's state machines might become confused if it sees an **ACK** without **AS***. It is guaranteed that if it is being accessed, **AS*** will be present. Otherwise it should neither know nor care that a cycle is going on.

Differences in RESET* Behavior

The state of the **RESET*** signal is not guaranteed when power is applied to the host. This is because it is difficult to guarantee the state of any electrical circuit until the power supply outputs are within their specified ranges. The designer of an SBus card should make no assumptions about the **RESET*** signal other than those shown in Figure 5.4.

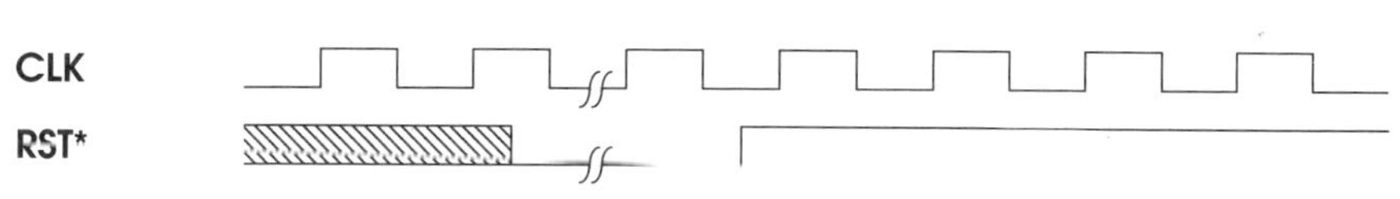

RST* must remain asserted (low) for at least 512 clocks after the power supplies are stable. De-assertion is synchronous.

FIGURE 5.4. *RST* Starts in an Unknown State.*

There are several variations of **RESET*** behavior that are legal and possible. One of these is shown in Figure 5.5. In this case the **RESET*** signal has drifted high with the power supply and appears to be asserted at first. This case is characteristic of the SPARCstation 1/1+ family of machines. Some early add-on cards were originally designed in such a way that their internal state-machines became confused if this initial high-to-low transition did not occur.

There are a number of reasons why it is important not to make any assumptions about the initial state of the **RESET*** line. There is no guarantee, for example, that the **RESET*** line will ever reach a valid logic level (high or low) during the time its state is unknown. Any transition that does occur is not guaranteed to be synchronous, either. It may occur at any time and could cause metastability in the state-machines that receive it.

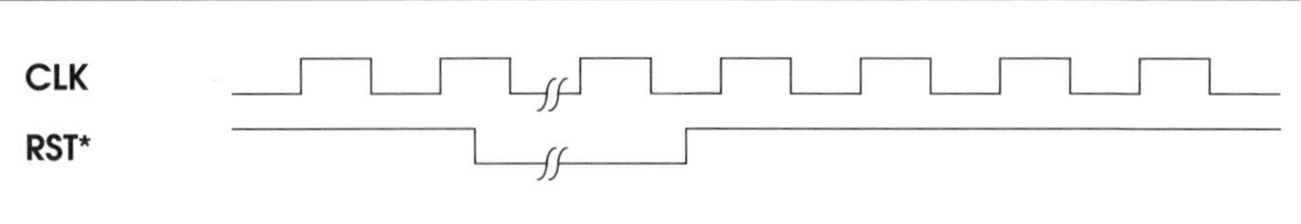

FIGURE 5.5. *RST* May Start De-asserted (High).*

Most importantly, however, there is no guarantee that *any* transition will occur! Another perfectly valid variant of **RESET*** behavior is shown in Figure 5.6. In this case the **RESET*** line starts and stays low until after the requisite 512 clocks occur. This variation is characteristic of the SPARCstation 2 family of machines. This variation is actually preferable to the first, because it reduces the chances that an SBus device will power-up in an unusual state and then try to do anything *before* it has been reset (if you are building a host and wish to guarantee this behavior, investigate using a depletion-mode FET transistor as a clamp; these devices will sink current until actively turned *off*).

Differences in Time from RESET* to First Probing

There is no minimum time limit between the de-assertion of **RESET*** and the first access made to any SBus device. The time that does elapse may vary substantially, and depends on both the type of host and the slot occupied. Generally speaking, any host will keep this time as small as possible, to reduce the overall time required to boot the machine.

Unfortunately, this can cause problems for some cards which require extensive initialization time after **RESET***. Cards with local processors are one possible example. Cards that contain programmable logic that must be loaded at power-up, such as XILINX FPGAs, are also at risk.

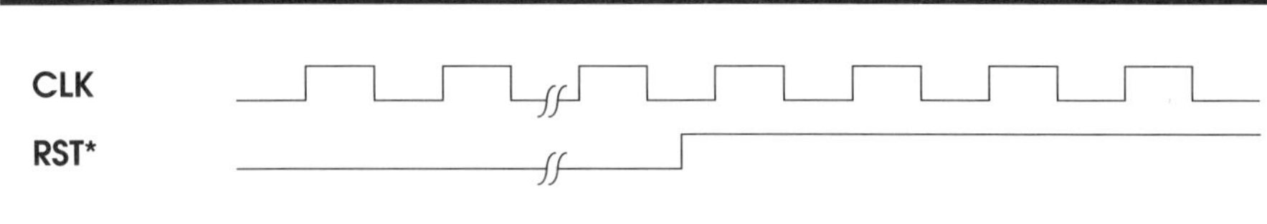

FIGURE 5.6. *RST* May Start Asserted (Low).*

Designers of such cards must guarantee that enough logic and intelligence is hard-wired into their design to allow it to respond to any access immediately after **RESET*** is de-asserted. If this happens before the card has adequately initialized itself, then it is acceptable to issue rerun acknowledgments to such accesses until initialization is complete.

Interrupt De-Assertion Might be Prolonged

The interrupt lines are shared, open-collector signals pulled up by 10 Kohm resistors. When no longer driven, only the resistor pulls the signal back toward the de-asserted state. Rise times of two microseconds or more are not uncommon, and high-speed RISC processors can execute dozens of instructions in this amount of time.

It's entirely possible that an interrupt might be serviced, but that the interrupt signal cannot de-assert quickly enough to prevent it from being recognized again. Several SBus driver developers have reported encountering this problem.

The author is unaware of any technically elegant solution to this problem as of this writing. One thing that driver writers can do to limit their exposure is to cause their device to release its interrupt request as soon as possible, so that it can get a head start on the way up while the rest of the interrupt service routine does its job. A longer-perspective solution may require that future SBus hosts reduce the value of the **IRQ(7:1)*** pull-up resistors from 10 KOhm to 2KOhm, however, and so SBus hardware designers should choose their interrupt drivers accordingly.

Don't Depend on Translation Cycles

Most SBus transfers start with a translation cycle, but this isn't always the case. In some hosts the central processor has a 'private' path to the memory management unit that performs SBus DVMA translations. This allows the translation to occur in parallel to other SBus activity, instead of in series with it. This saves transaction overhead on host-initiated transfers, boosting the overall level of performance. Sun Microsystems' SPARCstation 1+ is one example of a machine which does this.

The obvious implication here is that a slave access may begin immediately, without the effective 'warning' that a preceding translation cycle might otherwise give you. The physical address lines, **SIZ**, and **READ** may all be asserted at the same time, just before the clock cycle in which **AS*** is asserted.

This should not be a problem; all these signals will meet their setup time requirements. One caveat, though, is that a casual glance at SBus timing diagrams might lead you to believe that you have one or more cycles of additional setup time on **SIZ** and **READ**. Don't count on it.

Another possible complication is that when bus sizing occurs, the follow-on cycles can happen very quickly. In most cases **AS*** will only be de-asserted for the single clock immediately after the acknowledgment is sensed (**SEL*** may not de-assert at all). If a slave's state machines don't look for and make note of this, they will not be able to discern one cycle from the next. In that case the slave would see only one apparent transfer that takes an unusually long period of time to de-assert **AS***. The master will see the successful completion of the first transfer (with an acknowledgment indicating that bus sizing is necessary), but the first follow-on cycle will time-out.

Drive ACK signals Only When Necessary

An SBus card should not enable its **ACK** line drivers whenever selected (**AS*** and **SEL*** are asserted). This may at first seem a simple and logical thing to do, but there can be drastic consequences.

Consider what would happen if a slave device such as this never completes the access, due to some software or hardware failure. The SBus controller would eventually step in to end the cycle with an error acknowledgment. If the slave is already driving the **ACK** lines, though, there will be a bus fight. Because the SBus Controller's **ACK** line drivers are commonly relatively weak ASIC outputs, this is one fight the controller probably can't win. The result will be a system that 'hangs', stopping dead in its tracks.

It is only appropriate to drive the **ACK** lines during the cycle in which the acknowledgment is given, and on the subsequent cycle during which the lines must be actively driven to the idle state.

Write (Non) Time-out Bug

SPARCstation 1-class machines contain a hardware bug which prevents them from generating a time-out on write operations. Instead the machine will completely cease to function and all SBus activity will stop in its tracks. Only by being reset (in hardware) or by being power-cycled will the machine recover.

Each designer will have to decide how best to handle this situation within the constraints of his or her design. Complicating the matter is that there is a related bug in early Open Boot Proms (see that discussion starting on page 183). One key element, though, is that writes within your slave's device space should never be ignored; they must be acknowledged.

Rerun Ack Bug

The specification requires that a master which receives a rerun acknowledgment must relinquish the bus and try the transfer again later, *after* re-arbitrating for bus control. This is important, because one reason the rerun acknowledge may have been issued is that another transfer, perhaps initiated by a different master, must complete first.

The CPU master in a SPARCstation 1/1+ class machine does not react as required, however. Instead of releasing the bus and letting another master in, it retains bus ownership and repeatedly retries the same operation until it completes, or until 256 consecutive attempts have produced the same result.

Minimize Address Space Usage

The SBus provides a very large physical address space to each slot, even in hosts which provide only the 25-bit address subset. Still, it is a good idea to limit the address space your card needs wherever possible, and to concentrate this in the low end of the address range.

Naturally, designers seek to limit the amount of address decoding necessary. This includes limiting both the number of address lines that must be decoded, and the logic required to do it. This simplifies the design and may reduce its cost. When given a very large address space this tends to result in a design that breaks up the address space into multiple equal-sized pieces which are at least big enough for the largest addressable element.

For example, consider the case of an SBus slave that contains a 32 Kbyte ID PROM, a two byte status register, a 128 Kbyte buffer array, and a 64 byte FIFO. The simplest address decoding scheme for this situation is a 2-4 decoder which decodes **PA(18:17)**, producing four spaces of 128 Kbytes each. This is big enough for the buffer, and more than adequate for each of the other elements. Much of the address space carved out isn't actually needed, and there are (sometimes big) holes in the address map, but only a fraction of the space available is used. There is little incentive to spend design resources to recover address space when there is a surplus on hand, right?

In this case and others that may be true, but there are other factors to consider. The first and most obvious is that there is strong incentive not to *require* any more space than the 32 Mbytes which can be addresses per slot in machines which only drive **PA(24:0)**. Otherwise, parts of that space will not be accessible in such machines. It is perfectly reasonable, of course, to design the card in such a way that it can make use of all 28 physical address bits, *if available*, as long as the card will also function when only 25 bits are available. Some bus-bridges may provide even fewer than 25 physical address bits, and so SBus cards should generally concentrate their address space usage in the lowest part of the space available.

Limits on memory mapping resources provide further incentives for reducing address space usage as much as possible. The system's mapping resources must be shared by all devices, and overall performance will be affected by how efficiently these resources are used. Also, an SBus card will only be able to map at most 2-4 Mbytes of address space at any one time. If the card's address space extends beyond this then its driver will need to dynamically re-map its resources as necessary. Bus-bridges may limit this even further, because they might also contain limited mapping resources.

Use the READ Signal Carefully

The SBus' **Read** signal is used to indicate the direction of a transfer on the SBus. When the signal is active it indicates that a read is being performed; i.e. that data is being transferred from the slave to the master. The **Read** signal is a level, not a strobe. It indicates that a read is occurring, but it does not contain any information about exactly *when* the read occurs. This must be derived from other sources.

It follows that when **Read** is not asserted, a write operation will occur (data is transferred from the master to the slave). As in the previous case, though, no indication is included about the timing of the transfer.

Let's consider now how we will interface the **Read** signal to our logic. A sample schematic diagram for one part of a possible SBus interface is shown in Figure 5.7. For simplicity's sake the data and other buffers necessary are not shown, and neither is the circuit that generates the acknowledgment (in this case 'byte') at the right time. **SEL*** is qualified with **AS*** and the result is sampled synchronously, as suggested in the previous section. The resulting signal is used to enable a simple address decoder. This, in turn provides a an active low chip select to a random-access memory array. This memory also has a bidirectional data bus, which we have connected to the high order byte on the SBus, and a **Read/Write*** pin which we have connected to **Read*** (ostensibly, this seems the obvious thing to do).

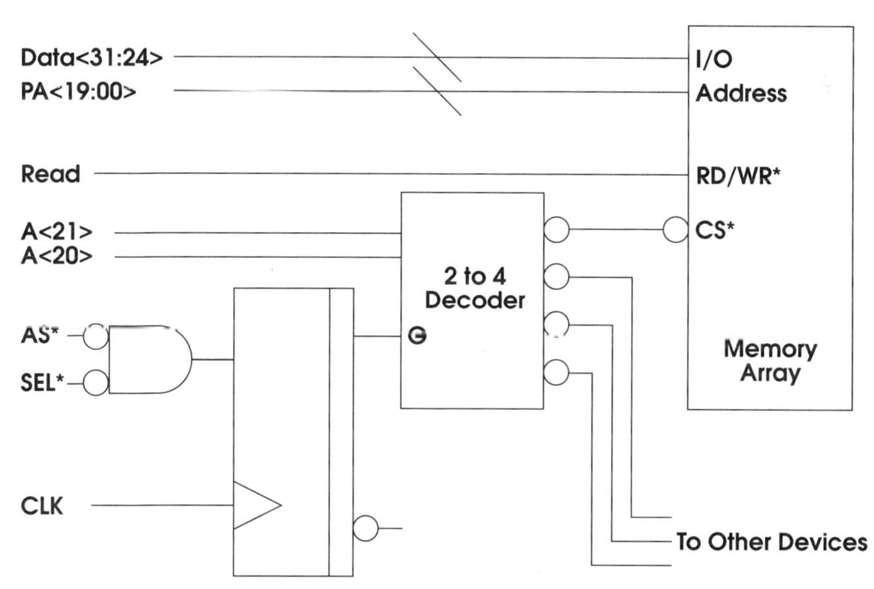

FIGURE 5.7. *Sample SBus Interface Circuit with Problems.*

A timing diagram for read operations is shown in Figure 5.8. Clearly, this circuit works pretty well for reads. First, **AS* x SEL*** is asserted, and this is synchronized by the flip flop. The result is factored into **CS***, which becomes effectively a copy of **AS*** x **SEL*** delayed by one clock. **RD/WR*** is high, and is guaranteed to be stable before **AS*** is sampled low. It is also guaranteed to remain stable for at least one clock following the de-assertion of **AS***. These same relationships apply to the physical address lines.

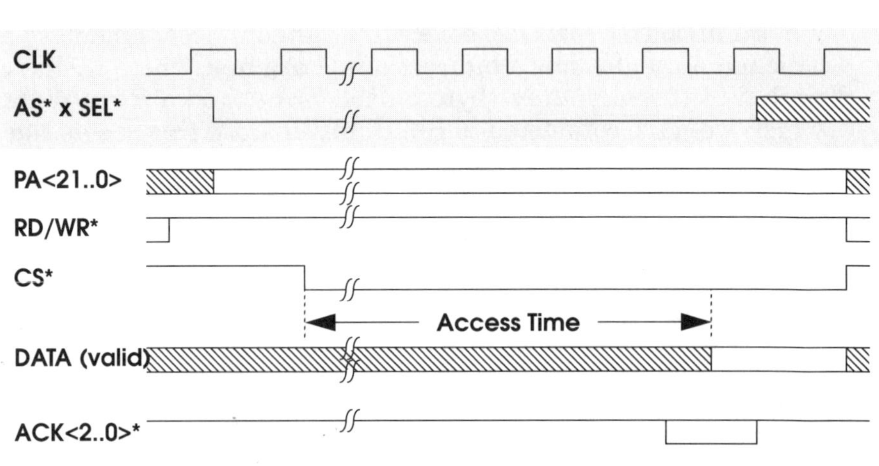

FIGURE 5.8. *Timing diagram for sample SBus interface circuit (Reads).*

The memory array now has a stable address, **RD/WR*** line, and **CS***, and it will commence the requested access. After some access time the valid data will appear on its I/O lines. The acknowledge signal then indicates that the slave is ready to transfer the data to the SBus master (in this case the acknowledge is timed so that it is sampled just after the data becomes valid, for best performance). SBus read data is transferred on the clock following the acknowledge. The physical address, **RD/WR*** line, and **CS*** remain stable throughout this cycle, and so the data will remain valid. It works! (For those of you who like to solve puzzles, there is at least one potential problem with this analysis so far that would bear further analysis in some very specialized applications. I'll provide the answer later in this section.)

Write operations do not work, however, which is easily seen in the timing diagram shown in Figure 5.9. The first part of the cycle actually looks promising. The **RD/WR*** signal can be asserted before the address is stable and the **CS*** is asserted. This might cause problems for some applications, but in most cases signals that indicate transfer direction are ignored unless the appropriate device select is also asserted. By the time **CS*** is asserted, the data, address and **RD/WR*** signals are valid and stable. After the write access time of the target device is satisfied, an acknowledgment is provided to indicate that the write operation is complete.

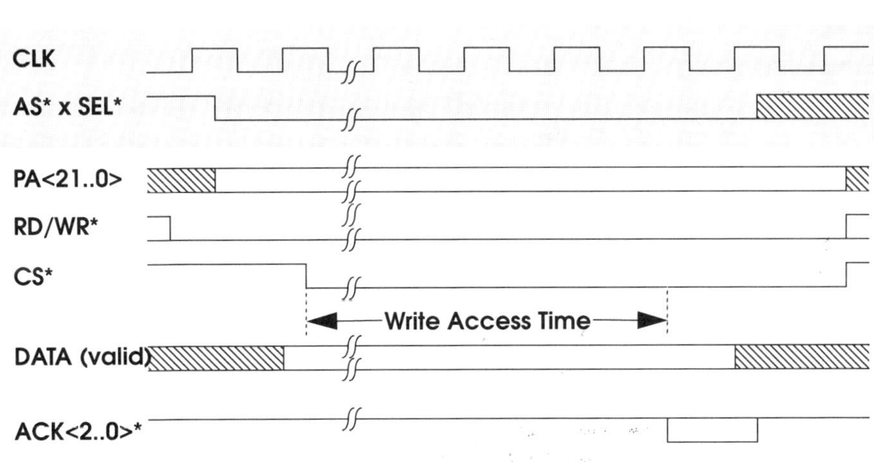

FIGURE 5.9. *Timing for sample SBus interface circuit (Writes).*

Here is where the problem lies. By the time the SBus master samples the acknowledge, the data has successfully been written into the memory array. Unfortunately the cycle does not end here. Having sampled the asserted acknowledge, the master is allowed to release or change the data lines. Since 0 nanoseconds of hold time is guaranteed, the data could change in the same instant as the high-to-low transition of the clock. **CS*** is still active, though, and **RD/WR*** is still signalling a write operation. It's highly likely that the desired write data will be corrupted or replaced by the "new" data.

What follows does not read like a methodical design approach, and it isn't. It's what happens in the lab with blue wires and soldering irons and extra PALs when a deadline approaches, and the design wasn't as methodical as it could have been.

A first attempt at a solution might be to include logic that de-asserts **CS*** upon assertion of an acknowledgment. If this is done in all cases, though, it will "break" read operations (look again at Figure 5.8. and convince yourself that this is true). If we complicate things even further so that we only de-assert **CS*** early for writes, we still haven't necessarily solved the problem. The write data may change faster than we can de-assert **CS***, and hold times could be easily violated. Ok, so pipeline the acknowledgment one level, effectively adding one wait-state to the transfer. Doing this it would be easy to guarantee our write data hold time, but it has cost us added complexity and we've lost performance in the bargain. It is possible to balance delays, or add asynchronous gates, or play other games that will solve these problems. If you do, though, you are likely to end up with a design that is sensitive and glitch laden.

Incidentally, similar arguments apply to de-activating the write by changing state on **RD/WR***, especially if inadvertent reads are undesirable (these may be dangerous if there are register bits nearby that can toggle when read, or if your logic is sensitive to minimum pulse widths).

An SBus design really shouldn't be this hard, and everyone's lives will be easier if you keep it synchronous. The state diagram for a synchronous state-machine that will do the job is shown in Figure 5.10. This diagram lacks only the transitions that would bring us back to idle from any illegal states, upon reset, if **AS*** goes away prematurely, and so on. The number of wait-states that can be inserted is variable, and can be as few as 0. If there are 0 wait-states, this interface is as fast as possible for reads because the acknowledgment will be asserted the cycle immediately following the one in which **AS*** is asserted. There is one extra state in the write path, though, which is required to guarantee write data hold time and solve the problems that have been discussed in this section.

Notice the transitions that keep this state-machine in the final read or write states while **AS*** remains active. These transitions exist to guarantee that the slave interface does not mistakenly begin another access if **AS*** is not de-asserted in a timely manner (as discussed previously). **AS*** must be de-asserted for at least one clock before this slave can begin another access.

Two simple variations of our state machine exist. The first of these is shown in Figure 5.11. Here, the fork between the write and read halves of the state diagram has been delayed by one state. This adds a wait-state to all read accesses. If performance is not as critical in your design as gates or combinatorial terms in an FPGA or PLD, then this change may be beneficial for you.

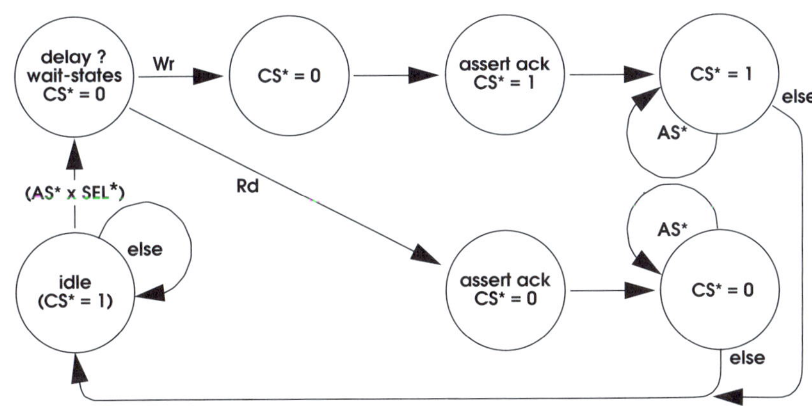

FIGURE 5.10. *State diagram for a solution to our sample interface problem.*

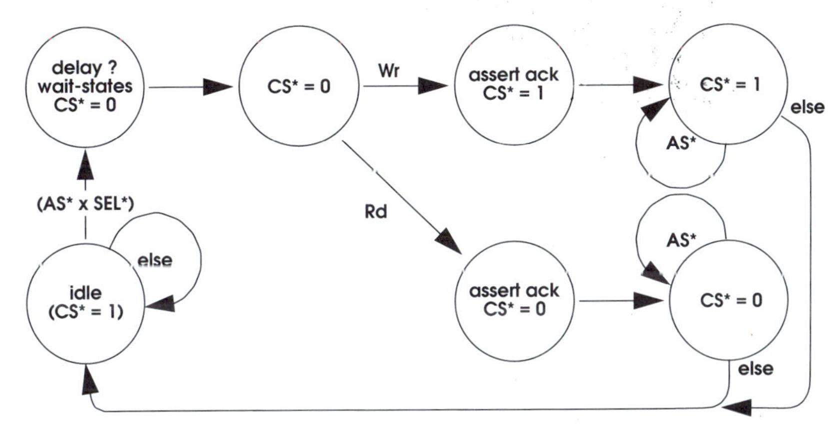

FIGURE 5.11. *Simplification is possible if one extra state on reads is acceptable.*

If, however, performance is of utmost importance, it is possible to remove the additional state from the write path, as shown in Figure 5.12. In this case the write data hold time must be guaranteed by pipelining (delaying by one clock through a register) the data.

Both write and read access times are symmetrical, now, and in the 0 wait-state case occur at the maximum rate possible.

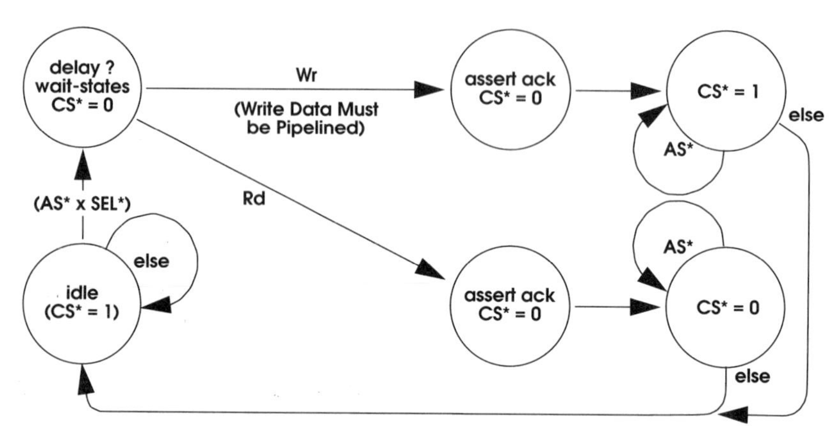

FIGURE 5.12. *Performance improvement is possible if write data is pipelined.*

Have you found the potential problem with read transfers for the sample circuit shown earlier in this section? Refer again to the timing diagram shown in Figure 5.8. The hold time guaranteed on the physical address lines and **Read** line (connected to the memory array's **RD/WR*** signal) is only 0 nanoseconds. Effectively, they could change at the very same instant as the high-to-low transition of the clock. There is a finite delay, though, through our synchronizing flip-flop and the combinatorial logic that follows it. As a result, it's possible that there could be an address change while **CS*** is still active, or while the decoder's enable is still active (which may cause one of the other device selects to glitch). This might cause an inadvertent read to begin. Even worse than that, if the **Read** signal changes fast enough an inadvertent write might begin! The actual application, the technology used, and any other on-board delays all factor into whether this is a real problem in each case, and if so, what to do about it.

Should SIZ, RD, PA, and DATA be Sampled Synchronously?

There is an observation on page 46 of SBus Specification B.0 which asserts that simple slaves do not need to sample (latch or register) the **SIZ**, **Read**, **Data**, or physical address lines. This is meant to reduce the amount of logic and the design complexity that SBus slaves must have.

There are situations, however, when it may be advantageous to sample some of or all of these signals. For example, it may be beneficial to sample the write data in order to guarantee data hold time (as discussed in the previous section) without incurring the extra cycle penalty this might otherwise require. Another situation is for Extended Mode (64-bit) transfers, when many of these signals must be sampled because they are multiplexed (see section 3.4.6 on page 74 for more information).

Generally speaking though, it will not be necessary for non-64-bit slaves to sample **SIZ**, **Read**, or the physical address lines. These signals are valid before **AS*** is asserted and remain stable until after **AS*** is de-asserted, which is sufficient for most applications. The DATA lines will be held stable on writes until after a suitable ACK is sampled, too. If for performance or other reasons any of these signals are sampled, though, then do so synchronously (as discussed in previous sections).

Of course, the fact that these signals are not *required* to be sampled in most cases makes it easier to sample them if you *choose* to. Setup times, hold times, and clocking are all much less critical than they otherwise would be.

Achievable Bus Bandwidth Depends on the Efficiency of all System Elements

The SBus is capable of very high transfer rates. The actual transfer rate achieved by a device is a function of three primary factors. The first of these is the host machine. The efficiency of the device itself is another factor. And the efficiency of all other devices is a factor as well.

The host determines the SBus clock rate, the number of cycles required to perform address translations, the access time of system memory, and so on. These and other architecturally dependent issues will define the ultimate SBus bandwidth that the host can support.

The device itself is an important consideration as well. Its support of DVMA and burst sizes can greatly affect its efficiency, as will the number of wait-states required for an access. Frequent rerun acknowledgments will reduce overall throughput, as will narrow port widths. All else being equal, an 8-bit wide device will have only 25% the efficiency of a 32-bit wide device.

The efficiency of other devices is critical, too. The SBus' bandwidth is shared among those devices attached to it. If one device is inefficient it reduces the overall bandwidth available to all.

A good illustration of this comes from the author's experiences as an SBus hardware applications engineer. One developer was surprised to find that his device was not able to get the bandwidth it needed to operate properly. His card was a simple printer interface with a relatively small data rate; something on the order of 10 kbytes per second. Even so, with his device, active underrun or overrun errors occasionally occurred. More commonly though, the host's ethernet interface reported errors which seemed to indicate bandwidth starvation.

Further investigation revealed that this developer's device was byte-wide. It was also relatively slow; large numbers of wait-states were inserted in each access. Even worse, his device-driver determined when a character needed to be read or written by polling a status register on the card. This status register was also only 8-bits wide, and required a large number of additional wait-states per access. For every byte transferred to or from this device, dozens of slow and inefficient SBus transfers were necessary. This device and the SBus it was attached to were literally "going nowhere fast!"

It is wise to make your SBus card as efficient as possible, even if blazing performance isn't needed in your application. This is like derating electrical components; there is no reason to push the limits if you don't have to, and leaving some "breathing room" will improve the reliability of your device and the overall system.

Ordering of Operations Should Be Considered

If building a host, it will be beneficial to spend a few moments considering the order in which operations are performed. Whenever possible, operations should be performed in the same order in which they were initiated. This seems obvious, but accomplishing it is not trivial in all cases.

Consider some host architecture in which the SBus is an offshoot of another, higher level interface. This other interface might be MBus, Futurebus+, VME, or almost any other architecture. Further, assume that the logic which connects the two contains a buffer or cache to maximize performance. This is a good solution in many cases because it reduces the necessity of one bus to "wait" for the other, and overall performance improves.

Depending on the design, however, this buffer function may upset the order of operations. If multiple buffers (or cache lines) are available and the sequence was not recorded, then the buffers might not be flushed in the same order they were filled.

Interrupts complicate the scenario, too. Suppose that a master interrupts the CPU as soon as it is aware that a write transfer has completed. It's entirely likely (even probable) that the transfer will complete on the SBus side before it completes on the side to which system memory is attached. If the latency is long enough, the CPU might receive the interrupt *before* the data gets written into memory, not after (as was the original intent). Some higher-level (software) mechanism may be necessary to prevent these interrupt "race" conditions from having undesirable effects.

5.3.2 Bus Sizing Issues

Bus sizing is a useful mechanism which allows masters and slaves to communicate without knowing up-front what data path width is appropriate. There are some issues of to be wary of, though, some of which are discussed in this section.

Bus Sizing Follow-on Transfers Are Not Guaranteed to Be Consecutive

When bus sizing occurs, a wide SBus transaction is broken up into several narrower ("byte-sized"?) pieces. Each of these *follow-on transfers* is a separate, independent transaction, and the slave need have no concept that one is related to any other. Because these transactions are completely separate, there is no guarantee that they will occur consecutively. In fact, it's entirely possible that another master could get in and perform a transfer with this slave before all segments of the original transfer have completed.

While this allows great flexibility and simplifies the design of the slave, it can cause difficulties. Consider the case in which a master asks for a 32 bit operation from an 8-bit slave. If the master supports bus sizing this will be broken up into four separate transfers of one byte each. The first byte is transferred in response to the original transfer request, and the rest are follow-on cycles. By the time the slave has provided the first byte, however, the master has already released **BR***. If it then wishes to do bus sizing and perform the necessary follow-on bus cycles, it must re-arbitrate for the bus. There is no guarantee that it will immediately regain control.

These transfers aren't necessarily contiguous, then, and another master could get in before the first master completes its transfer. This other master might even modify some or all of the same data that the first master is accessing. By the time that both operations complete, the data transferred to or from the first master may resemble neither the original word, nor the newly modified version. It might contain elements from both! The situation gets really ugly if the requests from both masters get bus-sized and interleaved.

The lesson here is to be careful if multiple masters are to access the same address spaces of any device that can cause bus sizing. Some form of arbitration or resource-locking may be necessary at a higher level; in the drivers via a main-memory semaphore, for example. Note that SBus atomic operations cannot be used for such a mechanism, because atomicity cannot be guaranteed on operations that perform bus sizing, for this very reason.

Follow-on Transfers Are Not Guaranteed

SBus masters are not required to support bus-sizing. Therefore follow-on transfers are not guaranteed to occur. Even if some follow-on transfers do occur, the master may still abandon the transaction whenever it chooses and drop all remaining follow-on transfers.

This should not usually pose any difficulties for the slave, which should be designed in such a way that it treats all transfers (including follow-on transfers) as discrete, independent transactions.

Follow-on Transfers Might Occur Very Quickly

When follow-on transfers do occur they may do so with very little delay. This is because only the low order address bits will usually be changing, and it often will not be necessary for a CPU master to perform any additional address translation. **AS*** may be de-asserted for only the single cycle immediately after the **Ack** code is recognized, as shown in Figure 5.13. This is the same cycle during which the slave will be sourcing data (if a read operation is in progress), and driving the **Ack** lines into the "off" state. If the slave does not sample **AS*** at this time it may mistakenly conclude that it has never been de-asserted. In that case it would wait patiently for the end of the current transfer, when in fact that transfer has ended and another has begun. Only the time-out which must inevitably occur will break the deadlock. To prevent this scenario, it is important that slaves should sample **AS*** at any time that it may change.

Notice that the **Sel*** is not necessarily de-asserted between follow-on cycles. This is one reason it is so important to qualify **Sel*** with **AS***.

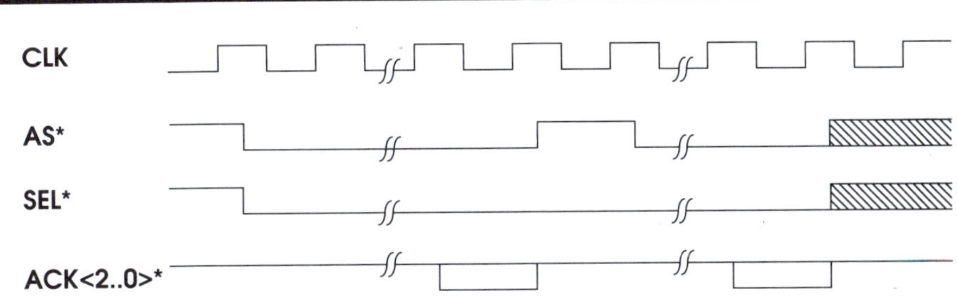

FIGURE 5.13. *One Example of Follow-on Transfer Timing.*

5.3.3 Burst Operation Issues

Generally speaking, burst operations are simple extensions of the SBus protocol, with very few differences between the multi-word burst and the single word, half-word, or byte equivalent. A few of those issues are discussed here.

On Bursts, How Does the Address Change?

On burst operations the slave is responsible for altering the physical address as necessary. For each word transferred the word address must be incremented in a manner consistent with burst address wrapping (as discussed in section 3.4.3). Note that the **PA(27:0)** signals do not actually change in any way; all this is internal to the slave.

Care must be taken to update this address at the right times in the transfer. The address must always remain stable through the cycle when data is transferred. This means that for write operations it must remain stable until the clock edge at which the associated **Ack*** code will be sampled. For read operations it must remain stable for at least two cycles longer; beyond the clock cycle which follows that in which **Ack*** is valid. Theoretically, the address could be updated here for write operations, but that adds at least two cycles to the transfer for each word; not a palatable alternative.

No generic "solution" to this issue exists. Each design will have its own unique capabilities and requirements. And with careful pipelining there may be no problem which needs solving!

Rerun Acknowledgments Are Only Allowed on the First Transfer

A slave may issue a rerun acknowledgment to a burst transfer request, but only on the first word of a transfer (or long-word for extended mode operations). If it does, then the burst transfer will stop without any subsequent data or acknowledgment requirements. If the slave does not issue a rerun acknowledgment at first, then the slave may only use word (long-word) or error acknowledgments for the remainder of the transfer.

The reason for this is that it may not be possible to rerun a transfer that has already been partially completed. Data that had been queued in a FIFO, for example, might be irretrievable. Bus bridges and expansion boxes may also buffer burst operations, to de-couple the buses and limit the amount of time that each must be tied up waiting for an operation on the other to complete. In such a case it might be impossible for the bus-bridge to pass on any late occurring rerun acknowledgment, because the originator may have already completed the transfer on its side of the bridge.

Error Ack During Bursts Bug

There is a bug in the Cache Gate Array chip used in the SPARCstation 1/1+, IPC, SPARCengine 1E, and all related machines. This chip provides much of the SBus controller's function for these machines. These systems are capable of performing 4 word (16 byte) bursts. During such a transfer, the SBus controller function expects to see exactly four corresponding word acknowledgments. It does not recognize error acknowledgments. If one occurs it is ignored by the controller (although the transfer's master will see it and know that an error occurred). The controller will wait, forever if necessary, until it sees four word acknowledgments on this transfer. It will not de-assert **AS*** until that happens, nor will it grant access to the bus to any other master.

The slave is only required to issue four *total* acknowledgments during this transfer. If one or more of those is an error acknowledgment then the controller will not see the four word acknowledgments that it wants.

The result is that the system stops dead in its tracks. In effect, its SBus is being held hostage while the controller demands its word acknowledgments as ransom. The time-out mechanism cannot come to the rescue here, because the error acknowledgments it issues are not recognized by the controller either! Only a hardware reset or a power-cycle will break the stalemate.

Machines of this type cannot generate burst operations via programmed I/O. Your board's slave interface can probably ignore this problem with clear conscience, unless there is a chance it will be involved in peer-to-peer transfers (from one SBus add-in to another, without the host's direct involvement). This is likely for frame buffers, but is otherwise limited to specialized applications.

Master interfaces cannot ignore this problem, though. Most SBus burst transfers in this class of machine are DVMA transfers targeted at system memory. In these hosts system memory can and does generate error acknowledgments during bursts if an error is detected. This might be a hardware failure in the host, but more commonly it is an attempted illegal access. In other words, usually the master (or the master's software driver) is the root cause.

Vigilant quality assurance and stringent testing under a variety of conditions can help insure that problems of this sort remain rare. No quality-assurance program can ever be perfect, however, and some hardware assistance in your master interface can help guarantee that any bugs that do fall through the cracks in testing can still be reported intelligently. Even a cryptic error message is better than a brain-dead machine.

Also consider simply "paying the ransom." Design your master so that it counts the word acknowledgments it receives on a burst request. If it is installed in a machine that contains this bug, then let it guarantee that the controller receives four total, even if it has to generate acknowledgments itself to make up any deficit! Such a master could tack on additional acknowledges until the bus is freed. The master knows that these are bogus acknowledges, of course, and the slave has finished the transfer and should no longer be driving the **ACK** lines if designed properly (still, it might be prudent to wait for some relatively long time to pass, to lessen the probability of a bus-fight on these lines).

The concept of an SBus master driving the ACK lines under any circumstances is unusual. Strictly speaking, it's probably not specification compliant either. Remember, though, that unless some drastic measure is taken, the machine is dead. "Defibrillating" it in this way may be the best solution.

These extra acks are not appropriate in any host which does not contain the bug in question. They would likely overlap and interfere with a subsequent cycle. Some test would be necessary, and the master's behavior adjusted based on the result. This could be tested in FCode at probe-time, or by the driver, or it could be a parameter set once during installation and then stored in EEPROM or NVRAM.

Burst Transfer Issues on Early L64853A SBus DMA Interface ASICS

Early versions (Revision A) of the LSI LOGIC L64853A SBus DMA Interface ASIC contain a bug related to burst transfers. This chip is a DVMA capable SBus interface, and the bug occurs when the L64853A receives a rerun acknowledgment to a DMA burst transfer it has initiated. The result is data which is lost or damaged. Apparently, for every rerun received on the operation, exactly one word of data is lost. This problem was probably not discovered right away because early SBus hosts never issued reruns on system memory accesses.

Once discovered, this bug was easily isolated and fixed. This part's Revision B resulted, which works properly in this regard. Few of the Revision A parts were shipped to vendors before this, and so the probability of encountering difficulties is remote.

To determine the revision level of one of these components, examine its markings, which should be of the following form:

```
LSI LOGIC
L64853AQC-25
DMA PLUS CTRL
<assembly lot number>
<date code>
<revision level>
<place of manufacture>
```

The first two or three digits of the revision level are important in this case. If these digits are "A8," then the component is a revision A part. If these digits are "0A8," then this is a revision B part. If neither of these codes is found then contact LSI Logic directly for more information.

5.3.4 Atomic Operation Issues

SBus atomic operations are designed primarily to support resource locking at a hardware level. The usefulness of this feature is unfortunately restricted by a number of caveats, restrictions, bugs, and configurability issues (especially with bus bridges and expansion boxes).

As a result, SBus designers are urged to investigate other options before designing support for atomic operations into your products. Semaphores in main memory may be a good choice. This allows resources to be managed at the driver level, and atomic operations within main memory are much simpler and less configuration dependent.

If atomic operations must be used then proceed cautiously. The information contained within this section will help.

'Atomic' Operations aren't Atomic!

Generally, the term *atomic operation* is used in reference to an event or transfer that can not be interrupted, or a failure will result. Usually this type of operation is used to gain exclusive access to a device, register, or the like. These are sometimes also called *locked* or *indivisible* operations.

An SBus 'atomic' operation does not do this, however. They cannot guarantee exclusive access to a resource. Instead, their purpose is to gain exclusive access to the SBus. This sounds like a subtle distinction, and for some architectures and some devices it is effectively the same thing. In most cases, though, the distinction is critical. If not understood, it will result in many inter operability problems. Future revisions of the SBus specification will rename these operations *Consecutive Transfers,* to help avoid confusion.

The distinction becomes clearer when single- and multi-ported devices are contrasted. A simple example of each is block diagramed in Figure 5.14. In the single ported case there is only one path in and out of the register, buffer, FIFO, or other part of the device which needs to be accessed atomically. In this case that path is the SBus, and exclusive access to the SBus does indeed provide exclusive access to that part of the device.

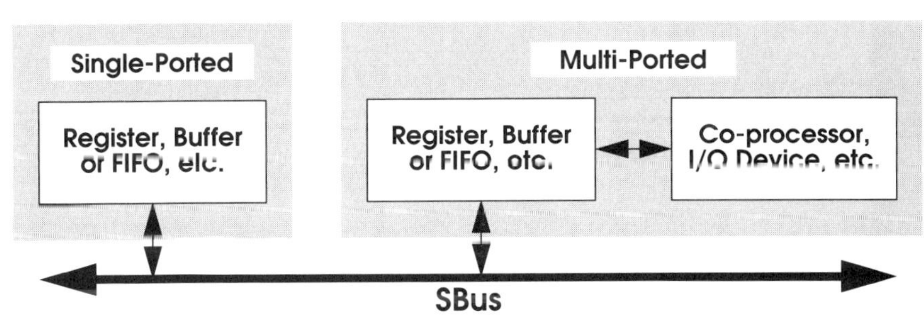

FIGURE 5.14. *Single-Ported and Multi-Ported Structures.*

The same is not true in the multi-ported case. Here, the part of the device which is being accessed can be reached by either of two paths. One is the SBus, as before. The other path connects to a co-processor, I/O device, or some similar function. This path allows that function to change the register's or FIFO's contents, whether or not exclusive access is being attempted via the SBus.

To correctly perform a true atomic operation, all other paths in and out of the operation's target would have to be blocked while the operation is in progress. For standard mode (32-bit) transfers this cannot be achieved purely in hardware, because there is no purely protocol-based way for an SBus slave to know that an atomic operation is in progress! It cannot see the **BR*** and **BG*** handshake between the master and controller, which is the only way an atomic operation is indicated. Even if it could, this would not provide the necessary information soon enough.

There are many ways for a slave to support true atomic operations, if designed properly. For example, the slave could set aside a portion of its address space for atomic operations. Any access to that space would automatically block all ports in a multi-ported structure. Another alternative is to build a small arbiter, which has a single-ported request register accessible only via the SBus. SBus atomics *can* guarantee exclusive access to this, and so a higher-level protocol can be implemented to inform a slave that an atomic transaction will follow, allowing the slave to lock the resource. When evaluating such an option, make sure to consider how and when the resource is *unlocked*. This may be particularly important in the case of a failure, where the atomic transaction has not been able to complete.

Masters cannot guarantee that an operation will truly be atomic, except by insuring that only atomic-capable slaves are accessed. Most masters communicate with system memory, which is generally *not* guaranteed to perform true atomic operations. This is because system memory will often be multi-ported; there may be dedicated paths to one or more host CPUs and co-processors, other independent I/O buses (VME, Futurebus+, another SBus, etc.), and so on. Such an architecture was shown in Figure 3.7. Exclusive access to an SBus does not guarantee exclusive access to the higher-level bus, and so cannot guarantee true atomic access to system memory.

It is possible that masters exist which mistakenly assume that the SBus is capable of initiating true atomic operations. This is because early SBus host architectures shared pins and signals between the SBus and system memory. This effectively single-ported their system memory, because it was impossible to access independently of the SBus. This result is a side-effect of this particular trade-off; it is not an SBus feature.

Avoid Atomics that will Result in Bus Sizing

Bus-sizing operations are not guaranteed to be atomic, as mentioned above. What if the operation starts out as an atomic operation though? What if the master does not de-assert **BR***? Will the operation then be atomic even if bus-sizing occurs?

Unfortunately this can not be guaranteed either. Any atomic operation that is bus-sized almost certain violates the read-[optional dummy reads]-write structure required of atomic operations. SBus controllers (and especially bus-bridges and expansion boxes) may not allow such operations.

Therefore, atomic operations that result in bus sizing are not guaranteed to remain atomic. In fact, they may result in errors. Atomic operations should only be attempted if the master knows beforehand that bus sizing will not occur.

Do not use Atomics to Guarantee SBus Access

Atomic cycles are meant to be used only to perform certain very specific operations, usually related to inter-process signalling and resource locking. Unfortunately, they can also be used to grab and hold the SBus for a short period of time, locking out other masters. This is not a valid use and should be avoided. This can be especially serious if a large number of operations are performed contiguously; SBus hardware does not enforce the maximum limits on the number of transactions performed in an atomic operation.

Using atomic operations in this manner unfairly divides the bus bandwidth, and can increase the latency seen by other masters to such an extent that they will function erratically or fail altogether. Your board may fail, too, if it violates the read-[optional dummy reads]-write structure required of atomic operations.

Atomic Operations, Bus Bridges, and Expansion Boxes

Atomic operations cause special problems with bus bridges and expansion boxes for a number of reasons. By their very nature, atomic operations must retain control of the SBus until completed. This can adversely affect performance (and especially bus latency) in even the best case, but bus bridges or expansion boxes make the situation much worse and more complicated.

In any bus bridge, it will often be necessary to arbitrate for the remote bus before an operation can be completed. This will add an indeterminate amount of time to the access. The bridge itself will usually also add delays because of the need to buffer signals, translate addresses, or add wait-states because of propagation delay in the interconnection. All these factors will negatively affect both transfer rate and latency.

Complicating matters, any operation that is atomic on the SBus side should also be atomic on the other side of the bus bridge as well. The other bus may not support atomic transactions, however, or it may only offer a different kind. Even if compatible atomic operations are possible (as in an SBus-SBus expansion box), determining when to do them can be problematic. The SBus does not provide the slave any indication that the operation is atomic! Not even the SBus controller can determine the start of an atomic operation at the beginning of the transfer.

These are just some of the complications that arise when atomic operations are used in an environment that contains bus bridges or expansion boxes. Atomic operations can be supported in most cases in such environments, but performance usually degrades and complexity often increases.

Beware of Hardware Bugs Related to Atomic Operations

SPARCstation 1/1+ class machines do not properly implement atomic operations. When some master other than the CPU master holds **BR*** asserted, indicating that an atomic operation *should* be in progress, the CPU master can still interrupt and insert intervening cycles. This is the result of a flaw in the design of the SBus controller implementations in these machines.

5.3.5 Split Transfer Issues

"Split" transfers are common in I/O bus architectures. They can be an advantage when an operation might take a long time to perform. Upon receiving a request, the slave begins to work on it, but does not require the master to wait for the result. Instead, it "disconnects," allowing other bus traffic to flow. Sometime later, after the slave has finished the requested operation, the master and slave will "re-connect" to transfer final status and data (if necessary).

The potential advantages of such an approach are clear; bus bandwidth and latency are not so adversely affected by devices with long access times. If the slave is multi-threaded (or contains independent devices), it may even be able to accept additional requests while working on the first.

Unfortunately, the SBus was not designed to explicitly support split transfers. Early on, though, it was decided that rerun acknowledgments could be used to accomplish the "disconnect." If some mechanism could be found to effect the "re-connect," too, then an important new feature could be used.

Toward this end, additional requirements were folded into early revisions of the SBus Specification. Recent work has shown that split transfers are much more complex than first thought, though. Some of the changes made to the specification on their behalf have unforeseen drawbacks. As a result, future revisions of the specification may modify and clarify behaviors related to split transfers. Until then, a technically sound split-transfer feature may be difficult to achieve in all but some very specialized circumstances. Caution is urged when considering their use.

What Does a Rerun Acknowledgment Mean?

At its simplest, a rerun acknowledgment is the way a slave tells the master to, "Go away." The slave does this when it is busy for the moment and does not want to be disturbed. For the purposes of this discussion we will call this a "state-less" rerun, because the slave does not attempt to remember anything about the transfer. Once the slave is able to accept transfers again, it can (and should) respond to any transfer request, in any order. If a master "abandons" a transfer (does not re-attempt a transfer that was given a rerun acknowledgment) the slave will neither realize this, nor care.

For split transfers, the meaning of a rerun acknowledgment changes subtly, to mean: "I'll start working on it. Come back later." This will be called a "state-full" rerun, because the slave explicitly must retain information about the desired operation (address, size, direction, etc.). In effect, it needs to attach some form of "claim check" to the transfer so that when the master does return, it is handed the proper status and (if a read) data. Any future access that does not present a valid "claim-check" will be re-buffed with a rerun acknowledgment. This might be a state-full rurun, too, if the slave can handle multiple requests and if there are resources available. Otherwise, it will be a state-less rerun.

Notice now that it might be important that the master does *not* abandon the transfer. If it did, then unclaimed transfers could clog the slave for indefinite periods of time.

Operation Abandonment and Multi-Tasking

To prevent a state-full slave from becoming clogged in this fashion, a requirement was added to the specification that forbids masters from abandoning transfers. A master must always re-attempt a transfer that has received a rerun acknowledgment.

A further requirement was added at the same time: not only must the master re-attempt the transfer, but it must not attempt any other transfer in the meantime. In other words, a master may only have one request in process (pending) at a given time. The intent of this requirement was to reduce the amount of state that a state-full slave must keep. With this requirement, simply knowing which master is performing the transfer is enough information to tag (attach a claim-check to) the transaction. It is not necessary to store and then compare addresses, sizes, and so on. Secondary benefits include a reduction in the number of requests that may be pending at any one time.

At the time these decisions were made, these requirements seemed a logical way to solve the problems, and they are simple and straightforward. What was not foreseen, though, were some of the unfortunate implications of these requirements.

First, imagine a case where the slave *does* somehow become clogged, either because it is broken, or because some master accessing it is broken or is not designed with the new rules in mind. This might result in an infinite rerun deadlock, which could soak up a lot of the bus' bandwidth. It might also prevent any master from accessing this slave. It might not be possible to break this deadlock either, because the card's slave port might give (state-less) retries whenever the master interface is busy (LSI Logic's L64853 and L64853A interface chips behave this way).

If the master involved happens to be the host master the situation is particularly problematic. The master can't abandon the transfer once given a rerun acknowledgment, and it can't do anything else either! So it can't do programmed I/O to investigate or clear the problem. The host's path to this bus is blocked. If the console device is somewhere on this bus, too, then the host can't even display an error message! An SBus reset may be the only choice, which destroys state information that would be useful to debug the problem. Besides, the slave might still be broken, and any access during probe or boot time could cause a repeat of the scenario, ad infinitum.

There are also cases when the master might be forced to treat a rerun acknowledgment as an error. This might happen if the master's latency expectations (and buffering capabilities) have been exceeded, and overrun or underrun errors are beginning to occur. In this case it does not benefit the master to continue attempting the transfer. In fact it may hamper error recovery mechanisms for the same reasons mentioned in the preceding paragraphs.

The architectural limitations that result from these requirements are perhaps the most troublesome, though. Suppose that the master is multi-tasking (multi-threaded). If an operation related to one of those tasks receives a rerun acknowledgment, then that task must be suspended until the operation can be completed. If this might take some time (the reason split transfers are used in the first place), then it seems logical to move on to something else in the short term. With the aforementioned restrictions, though, this can't be done if any SBus transfers are required other then the one still pending. Other tasks are blocked. This may result in substantial performance degradation for some applications.

A multi-tasking master may sound esoteric or unusually complex, but in reality it isn't. Any card which has more than one independent device could fall into this category. A combined ethernet and SCSI card is one common example. Another example would be a bridge to another bus, such as VME. The VME chassis might contain a number of masters, each of which represents one or more tasks. All those masters funnel their requests through a single SBus master, which is part of the bridge. That master may only have one request pending, though, so if any one of the requests gets a rerun acknowledgment for any reason, then the bridge is effectively blocked until that operation finally completes. None of the VME masters can access the SBus in the meantime.

Future revisions of the Specification will probably be modified so that some multi-tasking capabilities are provided. Masters still must not abandon transfers, but intervening operations may be allowed under certain circumstances. For backward compatibility this feature is one that must be explicitly (and programmatically) enabled. Default operation should always be single-tasking.

Controller "Blocking"

An issue closely related to split transfers is that of controller "blocking." This means that the controller can temporarily hold-off a master under certain circumstances. This might occur if the controller is aware of which slave a master is attempting to access, and is also aware that the slave isn't ready at the moment. Rather than "waste" an SBus transfer doomed to end in a rerun acknowledgment the controller might temporarily deny that master access to the bus.

Before a transfer begins, the controller usually *isn't* aware of which slave the master intends to access. If the master received a rerun acknowledgment on a previous transfer, though, then the controller *can* assume that the master will access the same slave again. Again, this is because revision B.0 only allows the master to have at most one request pending. The controller isn't privy to the internal state of most slaves. It might be, though, for the host's slave. It is capable of knowing, for example, exactly when this slave is ready to perform the last part of a split transfer.

Applied in this way, controller blocking is a good feature, though applicable only under special circumstances. The host slave and the controller can work together to maximize the efficiency of split transfers. There is another way that controller blocking can be applied, however, that must be done with more caution.

Revision B.0 of the specification allows the controller to intervene on all subsequent accesses to any slave that has generated a rerun acknowledgment. Only the master that received the rerun might be allowed to access that slave again, until a valid data acknowledgment ends the transfer. Most slaves cannot determine which master is attempting an access. This makes it difficult or impossible for the slave to adequately tag any transfer it wishes to split. If the controller behaves in the fashion described, though, the slave can be certain that if it gives a rerun acknowledgment, then the next access will be from the same master. Combine this with the restrictions that limit the master to one pending request, and prohibit it from abandoning transfers, and the result is a method which allows a slave to split a transfer without saving any state at all! The slave knows which master it's dealing with, and that the master will be re-attempting exactly the same transfer.

Unfortunately, the slave may not rely on this; it is not a guarantee. The controller is not required to behave in this fashion, and many (perhaps most) do not. Further, this type of controller blocking also has complications unforeseen at the time it was first included in the Specification.

The most obvious of these is that it may simply not be necessary. If the rerun acknowledgment is purely state-less, then the slave is unconcerned with which master and which transfer it next encounters. If the controller intervenes on behalf of this slave, it needlessly blocks accesses from other masters which the slave is perfectly capable of servicing. This impacts the bus' latency and overall performance. If the original master fails to re-attempt and complete the transfer for any reason, this block becomes permanent. This affects fault recovery.

Slaves with separate, independent devices or address spaces may suffer too. One device may be busy and issue a rerun acknowledgment when accessed, but another device on the same card could be idle at the same time. Similar arguments apply for slaves that have multiple resources, buffers, cache-lines, and so on. In either case, blocking access to these slaves by all but one master is another unnecessary action that may degrade performance.

As a result of these issues, future revisions of the specification may prohibit the controller from altering arbitration fairness because of rerun acknowledgments.

Performance Impacts

The purpose of split transfers is to enhance performance under circumstances where latency (or access time) is relatively long. Note that split-transfers have a built-in performance *penalty*, too, that must be balanced against their possible gains. This penalty is a result of the increased overhead associated with each transfer. After all, every split-transfer requires at least two bus transactions to complete. The first of these broadcasts the address, size, and transfer direction, and the last is the actual data transfer itself. This extra overhead reduces maximum bandwidths by about 10% to 30% or more. The actual penalty incurred depends on many factors, including the transfer size (large bursts are typically less affected), the number of transfers which receive rerun acknowledgments as a result, and so on.

Overall Recommendations

Split transfers can provide performance enhancements when used properly. They should be used only with caution, however, to insure maximum compatibility and the fewest possible architectural restrictions.

All slaves which split transfers should save as much information as possible about the transfer. This includes the address, the size, the direction, and if available, the master's identification. It should then compare this information with subsequent accesses to insure that the proper re-connect occurs.

Non-host slaves will generally not be able to distinguish between masters. A host slave may be able to (due to intimate connection with the controller). In either case, controllers should not limit access to slaves that have given a rerun acknowledgment. Slaves should not assume such "protection," either. In addition, slaves should not give rerun acknowledgments (make themselves inaccessible) whenever the master interfaces on the same card are waiting to perform a transfer.

5.4 Mechanical Issues

The following sections concentrate on issues related to the mechanical packaging and physical environment of the SBus.

Single-Piece Backplates Should Not Be Used

SBus Specification Revision A.2 introduced a modification to the design of the backplates used on SBus Cards. Previously, the design consisted only of a stamped metal plate which was shaped to mate with the SPARCstation 1's backpanel. This was then bolted to the SBus card or to a connector attached to the card.

The modified design consisted of a two-piece backplate which serves the same purpose. (See FIGURE 3.31. on page 103.) This change allows the backplate to be separated, resulting in a lower profile which is better suited to card-cage oriented applications such as VME and Futurebus+. This feature may also find use in some ultra-low-profile or very highly integrated desktop applications. The SPARCengine 1E and the SPARCserver 600 MP series are examples of products which already require the use of the double-piece backplate.

Despite the change in the specification, several vendors have maintained the single-piece backplate in their designs. The reasoning is usually based on the perception that the 2-piece backplate is a much more expensive custom fabrication, and that the size of the market which would require the 2-piece design is minuscule in comparison to the market served by the single piece design.

Both of these impressions were actually true just after the specification change was made, but that is changing rapidly. SBus volumes are high enough now that backplates are available off-the-shelf at low cost. Also, SBus is rapidly gaining favor with manufacturers of VME, Futurebus+, and other card-cage oriented systems, and that section of the market is likely to expand rapidly. These systems require the use of the 2-piece backplate, and entry into these markets won't be possible without it.

Some vendors realize this, and are hoping to convert their designs from a single-piece to a two-piece design when the time is right. Unfortunately, this slows the volume growth of the 2-piece designs and keeps everybody's long term costs up. This also leaves them with a large installed base of boards that can't take advantage of the new environments available to them.

IPC and IPX Mechanical Clearance Problems

The SPARCstation IPC and IPX use the same mechanical enclosure. This enclosure hinges like a clam-shell when opened, and is secured by two latches when in the closed position. These latches are on the sides of the enclosure, and one of them is very near one of the SBus slots. Early designs of this latch intruded on the volume set aside for SBus cards. The intrusion was small, but still posed a mechanical clearance problem for some SBus products.

The latches have been re-designed as a result of this problem, and incorporated into new machines shipped from the factory. This problem was caught early, and retro-fit kits were made available to owners of machines that were shipped before the change. Therefore it is unlikely that this problem will be encountered. If it is, though, contact Sun Microsystems for more information and the upgrade kit.

SPARCserver 600 MP Spring-finger Clearance Problems

When used on the SPARCserver 600 MP series, SBus cards are mounted on a VME form-factor processor board. These boards use "spring-fingers" (springy conductive metal strips) to seal the narrow row gaps between the VME boards in this system. This forms an EMI gasket, and helps attenuate the electromagnetic emissions that leak out of (or into) the enclosure.

When compressed, these spring-fingers can invade the volume reserved for SBus cards. This might cause a short circuit if an SBus card contains tall, conductive components near the backplate. The spring-fingers are electrically connected to the EMI enclosure, which is grounded. If the spring-fingers touch a component whose case is connected to one of the power rails (such as many transistors), then severe damage might result.

As this problem occurs only when the spring-fingers are compressed, it might not develop until after the processor card is re-installed in the system. Therefore, do not rely on eyes or ohmmeters alone to detect this.

To work around this problem, use an insulator. Refer to the 600 MP installation manuals for more information.

Surface-Mount Connector Compatibility

Surface mount SBus expansion connectors can be used in place of the through-hole variety, on either the motherboard or the add-on card. For mechanical strain relief these connectors will usually require some type of mounting tabs. These are not a problem in the case of the female connector used on the motherboard. There can be problems with the male connector used on the add-on card, however.

The SBus card outline has two holes which straddle the expansion connector. These holes are used for board-retention purposes; either an SBus retainer is installed in them, or snap-in or screw-in stand-offs can use them. The mounting tabs on a surface mount connector can interfere with these uses if improperly placed.

It is not enough that the holes in the connector's mounting tabs are aligned with the holes in the board. The tabs will still add depth that will make it difficult or impossible to attach a retainer. Also, the added depth will prevent the connector from being seated in hosts that use stand-offs. For example, boards which use a surface-mount connector with improperly placed mounting tabs cannot be used in a machine such as the SPARCserver 600 MP series, unless the stand-offs are removed. This is an undesirable workaround because the connector itself becomes the only means of retaining the board, and this is not sufficient protection against shock and vibration.

5.5 Firmware and Software Issues

There are many issues related to both firmware and software, most of which are far beyond the scope of both the SBus Specification and this book. Some key points may have relevance to the hardware design of an SBus product. Still others may impact the developers' efforts to debug their products. These issues are the focus of the following sections.

5.5.1 FCode and OBP Issues

This section concentrates on Forth, FCode, and Open Boot PROM related topics.

FCode Format Error in the SBus B.0 Specification

Revision B.0 of the SBus Specification incorrectly defines the FCode Program format. On page 114 of the specification there is a bulleted list in which the length and checksum entries are reversed. The correct order of the 8-byte program header is:

- "Magic" number (1 byte)
- Version number (1 byte)
- Checksum (2 bytes)
- Length (4 bytes)

"2.0 Pilot" Open Boot Proms

Early SPARCstation 2 machines were shipped with "pilot" versions of the Open Boot Prom. This version has a known problem with the **map-sbus** command, among others. To avoid difficulty, it should be replaced with a production version PROM.

To find out whether this is a problem on any machine, type **.version** at the OBP Toolkit's **ok** prompt. If the response begins "2.0 Pilot" then an upgraded PROM will be needed. If the response looks more like "2.0 Version <number and creation date>" then no change is necessary.

OBP Write Bug

Early versions of the Open Boot Prom have a bug in them that causes the system to crash if an error occurs on a write operation. This is true whether the slave issues an error acknowledgment, or if the cycle merely times out. Apparently, critical state on the OBP's stack is damaged when such an error occurs, and it is unable to continue intelligently. If you probe the machine it will appear to still be running, but it will either be unresponsive or insane. Errors on read operations do not cause these problems, but there is also a hardware error that will hang the machine when time-out errors occur on write transfers (that discussion starts on page 155).

A short term method which might help is to type "state-valid off" after every level 15 interrupt or L1-A key sequence. In the long run, though, one work-around worth considering is to always acknowledge writes with a valid, non-error code. This is not a technically elegant solution by any means. Suppose that a write is attempted to a read-only space in an SBus card (such as the EPROM space). It seems odd to acknowledge such a transfer at all, and especially with a code that is going to lead the master to believe the operation succeeded! Still, keep in mind that unexpected or illegal writes do not occur under normal circumstances. In most cases they result from "cockpit" or software errors early in debug.

If you are uncomfortable with losing this degree of error checking, consider putting in a "mode" option which allows illegal write operations to be ignored or acknowledged with errors. Turn this mode off for SPARCstation 1 family machines (determined by FCode at probe time perhaps), or after debug is completed.

Or report such errors either with interrupts or the **LateError*** signal.

Interrupt Mapping Bug

Versions 1.0 and 1.1 of the Open Boot Proms (primarily found on SPARCstation 1 machines) contain an interrupt mapping bug. SBus interrupt levels 6 and 7 were incorrectly mapped to CPU levels 9 and 13 respectively, instead of CPU levels 8 and 9. As a result, the **intr** property will be incorrect. If your card uses either SBus level 6 or 7, then you will need to make your customer aware that a version 1.2 or later Prom must be installed for the card to work properly.

"WFILL" and "LFILL" not Supported

"Fill" is an FCode that can be used to fill memory with an arbitrary byte value. "wfill" and "lfill" are word and long-word equivalents that are available in some early versions of the Open Boot Prom Toolkit. Due to PROM space limits, however, later versions of the toolkit support only the byte-wide "fill."

"Fill" is supported in all current versions of the toolkit and will be supported in all future versions. It optimally uses 8, 16, or 32 bit operations to perform the fill operation, and so is as efficient for purely filling or clearing operations as the wider variants. It can also start or end on any address, too (it does not need to be word or long-word aligned), and so is more flexible. The only real drawback is if the fill is not a simple pattern and really does require 16 or 32 bits. In this case, though, a simple loop can do the job, albeit less efficiently.

5.5.2 Software Matters

Discussions concerning software are inappropriate for this book to some degree, because fundamentally the SBus is not dependent on any particular architecture or operating system. The next few years are expected to see the SBus spreading far beyond just the Sun-compatible market.

As of this writing, though, most SBus applications are found on Sun-compatible workstations and servers. Most of these environments run some variant of SunOS or UNIX. Regardless of the SBus' eventual success in penetrating other segments of the computer industry, this market segment will likely always be an important one. For that reason most SBus developers who want to maximize their potential markets may benefit from some discussion of the software issues found in this chapter. An understanding of these concepts will help to maximize compatibility and minimize configuration dependent concerns.

Interrupt Service Latencies

UNIX is not a real-time operating system. It was not designed to be able to guarantee minimum amounts of processing time or maximum interrupt latencies. This last factor is particularly important for embedded-control applications, which often must be able to guarantee a timely response to an event. Interestingly, this doesn't necessarily mean that the response has to be fast, but only that it occur within a known period of time. For example, consider a device controller which must interrupt the CPU at the beginning of a transfer, and then buffer the incoming data until the interrupt is serviced. The amount of buffering this controller needs will be a function of the interrupt service latency. If this latency is not well bounded, then it will be difficult to determine how much buffering is needed. There may well be too much buffering most of the time (which results in a board that is usually more expensive and complex than necessary) and yet not enough in others (despite the usual overkill on buffering data is still occasionally lost).

By measurement, SunOS 4.XXc based machines have interrupt latencies that range from 20 to 100 milliseconds, with most responses at the lower end of the range. None of this can be guaranteed, though, because there are too many variables which factor into UNIX's interrupt response time.

Long term remedies are planned to meet the needs for real-time operating systems. SunOS version SVR4 is slated to include a "real-time" scheduler and a "pre-emptive" kernel, but as of this writing it is unclear how much this will help this situation. Also, Sun and Lynx Real-Time Systems have signed an agreement to port the latter's LynxOS (TM), a real-time UNIX implementation, to SPARC-compatible workstations and embedded board products. This version of LynxOS will be binary compatible with SVR4 SunOS. Applications can be developed in advance on Sun's developers' release of its SVR4 operating system and will run unmodified without recompilation when LynxOS is available.

In the short term, though, users who need guaranteed response times (or response times that are very short) should avoid involving UNIX directly whenever possible. Use sufficient local buffering. Make use of DVMA to transfer data directly to system memory (hardware latencies generally can be guaranteed, and are much shorter than interrupt service latencies). Set up "chains" for long transfers up-front, rather than on an as-needed when-needed basis. Consider using "mailbox" or "flag" memory locations to signal events back to applications programs or device drivers. Also consider local co- or embedded-processors when situations might require a decision "on-the-fly."

Interrupt Mappings Might Differ

SBus interrupts may be mapped (prioritized relative to other devices and tasks) differently from one machine to the next. This may cause incompatibilities if the interrupt levels aren't chosen carefully, and if the driver isn't written with this in mind. For more information on how to select the appropriate interrupt level, refer to the detailed discussion which begins on page 263.

There are a number of ways this can adversely affect the system if not taken into account beforehand. If the effective interrupt level in one machine is *lower* than the card was designed for, it may not get serviced frequently or quickly enough to prevent errors, such as overruns and underruns. If the effective interrupt level is *higher* than expected then this device is likely to work well, but some other device might not get the attention it deserves.

There are other interesting (translation: subtle and difficult to isolate) problems that can result from servicing an interrupt at too high a level. One example involves UNIX "STREAMS" drivers on SPARCserver 600 MP class machines. Much of the OS's STREAMS code effectively runs at SPARC level 10. If a STREAMS-based driver runs at a higher level than this, then there is a strong chance that some shared data structures may be corrupted (usually because they are used by the driver before the OS code has finished assembling them). The result is a "PANIC..." message in the console followed by a shutdown of the system.

On SPARCstation 1/1+ and 2/2+ class machines this problem cannot occur because even SBus interrupt level 7 (the highest priority) is mapped only to SPARC level 9. On SPARCserver class machines, however, both SBus interrupt levels 6 and 7 are mapped to SPARC levels higher than that at which the STREAMS OS code operates. These are mapped to SPARC levels 11 and 13 respectively.

Fortunately drivers can be written in such a way that they are relatively immune to differences in hardware mappings. Though these mappings and the SBus interrupt levels used by a card are usually hard-wired, the SPARC level that the driver executes at can be modified on the fly. This is done by using the hardware interrupt service routine only to schedule a software interrupt. In this way the hardware service routine executes quickly and then gets out of the way, while the actual work (and interface to the OS) is done in the software interrupt service routine at the level of your choice.

DVMA Size Limits

SBus hardware limits the maximum amount of data that can be moved in any single DVMA transfer to 64 bytes (or 128 bytes in the extended transfer mode). Obviously, though, a number of such transactions can be strung together in rapid succession to move very large amounts of data. Still, there are limits to the amount of data that can be transferred in this way. This is because the operating systems of most SBus based machines funnel the data through a DVMA array (or buffer). This ultimately limits the amount of memory that DVMA devices can map at any one time. The DVMA array is limited in size to about 760 Kbytes for typical SBus hosts. Generally, this amount can't be changed without extensive modifications to the OS' kernel.

The DVMA array is a shared resource. Ethernet, SCSI, and other DVMA devices also compete for its use. Even so, drivers which wish to perform large DVMA transfers should have little difficulty in allocating 400 or 500 Kbytes for this purpose. For transfers larger than this it will be necessary to dynamically allocate and de-allocate space. Under no circumstances should DVMA array space be statically allocated. That is unfair to other DVMA devices and is likely to seriously impact system performance.

Please also note that SunOS 4.0.3c (and derivatives) contain another historical limit which will constrain DVMA transfers to something less than 124 Kbytes in size. Fortunately this limit is not shared among all DVMA devices.

Modload Problems with Small Drivers

One vendor has submitted a bug-report concerning modload. Apparently modload doesn't work if the physical size of the object module is less than 8 Kbytes long.

This may be related to a previous problem discovered which can cause modload to fail if the length of the object module falls on a page boundary. For example, machines such as the SPARCstation 1 use pages which are 4 Kbytes deep. If the length of the object module being loaded is 4 Kbytes, 8 Kbytes, 12 Kbytes, etc., this bug may rear its ugly head.

Fortunately, there is a simple workaround for both these possible problems. Pad out the object module so that it is greater than 8 Kbytes long, and so that it does not fall on a page boundary. This last may sound difficult, because there is no fixed SBus page boundary size specified. Also, some machines may use reference MMUs, so that even one machine may contain a mix of several different page sizes. You are virtually guaranteed, though, that page sizes will always be a power of two. So pick an odd number that (just to be safe) is a few bytes distant from any of the more likely page boundaries.

DVMAMAP and RMINTR Routines Are Missing on SPARCengine 1E Family OS Releases

The OS kernel for the SPARCengine 1E family comes from the Sun4 source code branch, not the Sun4c branch followed by the SPARCstation 1/1+ and subsequent SBus based machines. As a result, the "dvmamap" and "rmintr" routines that many SBus drivers need are not available, and these drivers will not load.

A kernel patch is available that should eliminate this problem. Contact Sun Microsystems for more information.

Multiple Register Sets Bug

There is a bug in the SunOs Revision 4.1.1 kernel which will corrupt device tree structures for any card which declares multiple register sets. The sbus_decode_regprop() routine fails to increment an internal index while extracting information from the "reg" property.

A kernel patch is available from Sun Microsystems which corrects this problem.

Acknowledgments

XILINX is a trademark of Xilinx, Inc.

SunOS and Sun are trademarks of Sun Microsystems, Inc.

SPARC is a registered trademark of SPARC International, Inc.

UNIX is a registered trademark of UNIX System Laboratories, Inc. LynxOS is a trademark of Lynx Real-Time Systems, Inc.

References

Sun Microsystems, Inc., *SBulletin*, June 1990

Sun Microsystems, Inc., *SBulletin*, March 1991

Think Small and Low Power

6

The purpose of this chapter is to help designers work within the SBus form factors and power budgets. This might require a change in the way a design is approached, especially if most of the designer's experience is with larger, less integrated, backplane-oriented interfaces. Contrary to the initial impressions that such developers often have, it is quite possible to fit even very complex and sophisticated designs onto an SBus card.

This chapter will occasionally make its point by drawing examples from a real SBus product that the author helped design. This product is a single-width SBus slave card that contains two "fuzzy logic" coprocessors and all necessary support circuitry. The board is manufactured by a company that specializes in fuzzy logic design, and follows related VME- and ISA- (AT) based products that the company previously developed. This provides a good opportunity to examine the effort involved in "shrinking" a design to fit within the SBus' constraints. Particular attention will be paid to the contrasts between the SBus and ISA based designs.

A block diagram for this board (named the FCD10SBus Fuzzy Logic Accelerator) is shown in Figure 6.1. The board contains an SBus interface and ID PROM. The interrupt register shown allows the board's interrupt sources to be masked or tested. Certainly the most interesting parts are the fuzzy logic coprocessors themselves, highlighted in this diagram by the shaded areas. These coprocessors are identical to each other and fully independent. Each contains one ASIC, which is essentially a special-purpose microprocessor, and the "knowledge base" RAM, which serves as its control store. The instruction set of the microprocessor is optimized for evaluating fuzzy logic rules, and there are control registers and dual-ported RAM on the chip used for message passing and data transfer. The knowledge base of each coprocessor is built from 128 Kbytes of fast static RAM arranged 16 bits wide.

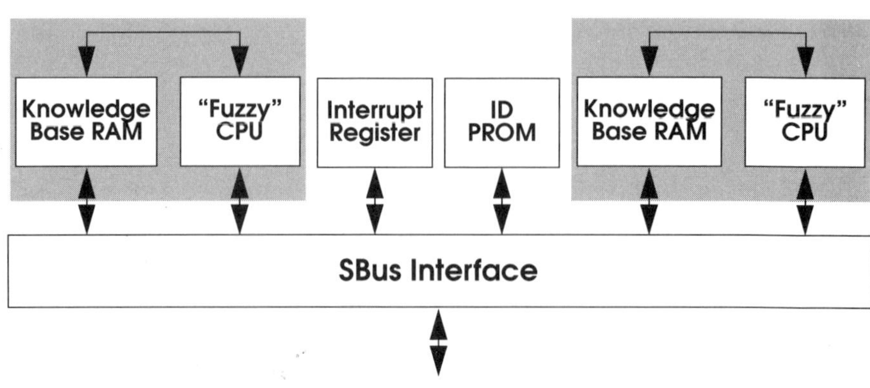

FIGURE 6.1. *Block Diagram of an SBus-Based Dual Fuzzy Logic Coprocessor.*

Address and data multiplexors are included which allow the RAM to be tested and downloaded from the host via the SBus.

While the emphasis of this chapter is on saving space and power, there are other benefits to be had as well. The initial reaction to much of the information presented here might be, "we can't do that, it costs too much." Some of the technologies or techniques are more expensive than what developers feel will fit within their budget. Or the reaction might be, "we don't have that kind of expertise, or the tools necessary." In some cases, more sophistication and experience is required. Or perhaps the feeling is, "we don't have the time, our schedules are very tight." Sometimes learning curves or lead times do seem uncomfortably long. When making any trade-off, though, it is important to consider its effect over the entire product life cycle, not just the development period.

For example, suppose that one option requires spending twice as much on the printed circuit board fabrication. That may seem drastic, but it probably isn't. Usually, the cost of a printed circuit board is only a small fraction of the device's manufacturing cost, and an even smaller fraction of the final selling price. Other factors might easily outweigh this additional cost. For example, perhaps money can be saved on other components (fewer or simpler parts might be needed). Perhaps the product can be brought to market faster, widening the window of opportunity (and profit). Or perhaps the customer will pay a higher margin for a more sophisticated product.

The manufacturing cost of the FCD10SBus card is just about twice that of the ISA-based card that preceded it, but the functionality has been doubled, too. Hence, the price/performance ratio has remained relatively constant. The increased cost mostly isn't because the technology used is more expensive, but simply because there is more of it. All this in a package almost half the size, and headed for a market that is much less cost-sensitive (hence profits are correspondingly higher).

6.1 Minimizing Space

One of the most commonly voiced concerns about the SBus is the perceived difficulty in fitting a design into the small form factors. It is often forgotten that those form factors are a key feature of the SBus, chosen specifically with high levels of integration in mind. It is possible to build even very complex designs within these constraints using technology readily available at moderate cost.

A variety of very high density packaging options are available for designers to use in shrinking a design onto the physically small SBus form-factors. There is sometimes a tendency to shy away from technologies such as these, out of fear that they are overly complex or expensive. This fear is all too often greatly exaggerated, because the technologies in question are no longer exotic, in fact, they are increasingly commonplace and cost-effective. It's important to remember, too, that the SBus was optimized with technologies such as this in mind. While the technology used may be more expensive, less of it is required to do the job.

For example, consider the case of moving an existing VME-based design built with through-hole technologies to the SBus. The simpler SBus interface and advances in technology since the VME board was designed will generally mean fewer, more power conservative components are necessary. This will save money a number of ways (reduced administration and inventory costs, higher manufacturing yields, etc.). The small SBus board will be cheaper and easier to manufacture than a VME card too (less fiberglass, fewer holes, no stiffener, smaller backplate, etc.). The savings made in these areas and others could offset or even surpass any additional costs incurred by switching to surface-mount technologies, or ASICs, or both.

The purpose of the following sections is to discuss strategies and techniques that will help SBus developers best use the board-area available to them. There are three basic themes woven throughout. First, use high density technologies. Second, make best use of the board area you've got, and finally, minimize board area requirements by making best use of the board's volume.

This last theme is less obvious than the first two. In areas where land is scarce or very expensive, real-estate developers often maximize their floor space while minimizing their land use by building tall, multistory structures. The skyscrapers found in almost every metropolitan area prove the effectiveness of this strategy, which works equally well for the SBus. While board area is often precious, the board volume is relatively generous in most cases.

6.1.1 Invest in ASICs and PALs

By far the best way to conserve SBus board area is to make good use of the ASIC and PLD technologies for which it was designed. Very complex circuitry can be implemented on single, small, inexpensive chips. Entire SBus interfaces have been implemented using only two or three such chips, a few passive resistors and capacitors, and a connector or two.

Off-The-Shelf Silicon

One of the easiest ways make best use of ASIC technology is to find a ready-made part which does the job for you. Off-the-shelf parts are often less expensive than a custom designed chip, because the development costs are spread across all companies that use it, instead of just the company that designs it. The parts are also available quicker, and with less risk (because the design is already substantially proven).

ASIC and VLSI components exist for virtually any function that the user might wish to incorporate on an SBus board. Sophisticated co-processors, DSP chips, graphics engines, and the like are all available in a wide variety of forms. This book will not address that, concentrating instead on the interface that any such component will need for use on the SBus. There are several off-the-shelf options for SBus interfaces available, and the selection is expected to grow as does the SBus' popularity.

LSI Logic Corporation has made a substantial investment in the SPARC and SBus silicon fields. This company produces a wide variety of chips and chip-sets which can be used to build SBus interfaces, SBus-to-MBus bridges, and even complete SPARC and SBus-based workstations. It is the SBus interfaces which are of primary interest here. Their first such product is the L64853 SBus DMA Controller. This part provides a complete 32-bit wide SBus interface capable of both master and slave operations. Devices may be interfaced via either of two independent ports; one 8-bits wide and the other 16-bits wide. This chip was optimized for Ethernet and SCSI interfaces, but it can also be adapted for other uses, sometimes with a small amount of additional logic. Internal byte packing and unpacking registers are used to minimize traffic on the SBus, too, by combining small data transfers so that most SBus operations can use its full 32-bit width. The L64853A is a more advanced version, because it can generate burst transfers. It is pin- and software-compatible with the L64853.

By the time of this book's publication, the "Goldchip" (MC92001), designed by Sun Microsystems, should also be available from Motorola. This chip provides a flexible, full-featured master and slave SBus interface capability, including burst and extended mode transfers. This part can support up to eight independent DMA channels, has a programmable wait-state generator and slave address decoder, and many other features which make it suitable for a vast array of SBus interface applications.

Custom Designed ASICs

Whenever buying "off-the-rack," whether it is a suit or an SBus interface ASIC, the fit probably won't be perfect. The suit's trousers might be a bit too long, and the coat's shoulders a bit too snug. The SBus interface ASIC might require some additional "glue" logic for your purpose, or it might not give you all the features that were hoped for. In some cases the benefits of getting something quickly and cheaply override the sacrifices that must be made. In other cases, though, it is more important that the fit is a good as possible. Then a tailored suit or a custom designed ASIC is more appropriate.

Many engineers (and their product managers) cringe at the thought of designing an ASIC. The impression is often that esoteric skills and a lot of time and money are required. Also, there is the ever-present risk that after all this work and time and money, the chip will contain some fatal flaw that will send you back to the drawing board for three months. Many people believe that designing ASICs is something that only big companies do, and that small companies haven't a chance.

A decade or so ago this was all mostly true. ASIC designs (including gate-arrays) did take a long time to complete. Logical simulation and timing analysis were often cumbersome and incomplete. Non-recurring engineering charges (NRE), which includes things like chip-layout and mask generation, were often tens of thousands of dollars. The lead-times required to get completed chips could be as long as 10 – 12 weeks, and if the chips weren't bought in a substantial quantity their price could be very steep.

Fortunately, it is much easier to build ASICs today, and even small companies with tight budgets and schedules can consider it. CAD tool technology has improved greatly, and designs can be done using higher level "languages" and standard function cells and macros. This means that the designer is not dealing with individual gates and transistors, but with functional blocks such as adders, or even entire SBus interface primitives! This reduces the amount of time necessary to do a design, and increases the probability of getting it right. Simulation and timing analysis tools have also improved, so that less work is necessary to provide greater coverage.

Many of these tools are used both for board and for chip design, so many companies may already have the basic CAD structure in place. Even if they do not, it still may not be necessary to acquire them. Many ASIC vendors provide design facilities that are fitted with an integrated set of design tools. Chip designers can visit these facilities and design their chips there. There are advantages to this kind of approach that go even beyond the fact that it is not necessary to buy the tools. Such design facilities are usually optimized and streamlined for that vendor's technology, and the interfaces are already established and debugged. Also, these facilities are usually staffed by people that are experienced in the development process, and can help guide the inexperienced designer.

Lead-times and costs have been reduced, too. Costs are very dependent on the vendor, the technology, and the complexity of the design, and so are difficult to discuss here. The reason that costs (both NRE and per-chip) have fallen, though, is that the number of ASICs being produced has mushroomed dramatically, and ASIC vendors can divide their overhead costs over a wider range. Their process lines have become more efficient, too, and typical lead times are now on the order of three to five weeks. Many companies also offer "HOTLOT" or "Red-Rush" accelerated schedules as options that can provide completed chips in days, not weeks.

Designing an ASIC is still not a trivial task, but it's not an impossible dream either. The cost and time associated with the design can be partially or completely offset, because complexity in the rest of the board can be reduced accordingly.

PALs and FPGAs

There are many types of Programmable Logic Devices (PLDs) which can be an attractive alternative to ASICs in many cases.

These include PAL (Programmable Array Logic) and FPGA (Field Programmable Gate Array) devices. These parts are programmed in much the same way EPROM devices are, often using the same equipment. Development tools are relatively inexpensive, and so are the parts themselves. PLDs provide customized logic, Like ASICs, but without the NRE costs and long lead-times. More importantly, design errors can easily be corrected, often within only minutes. This reduces the associated risks.

There are trade-offs, of course. PLDs generally use more power than ASICs, can be slower, have higher failure rates, and usually offer less logic and fewer pins. For low volumes they can be cost competitive, but as volumes increase they are more expensive than ASICs, too. To get around some of these issues, many PLD vendors provide the ability to "harden" a proven design, by producing a masked version from the original program. This allows the user to take advantage of both the short-term advantages of PLDs, and many of the long-term advantages of true ASICs.

CMOS PAL devices are well suited to SBus environments. One PAL and a small number of other devices can provide an entire 8-bit SBus slave interface, as shown in Figure 6.2. The peripheral could be almost any CMOS "microprocessor compatible" device, such as a serial port interface, digital-to-analog converter, analog-to-digital converter, DTMF encoder or decoder, RAM, FIFO, and so on.

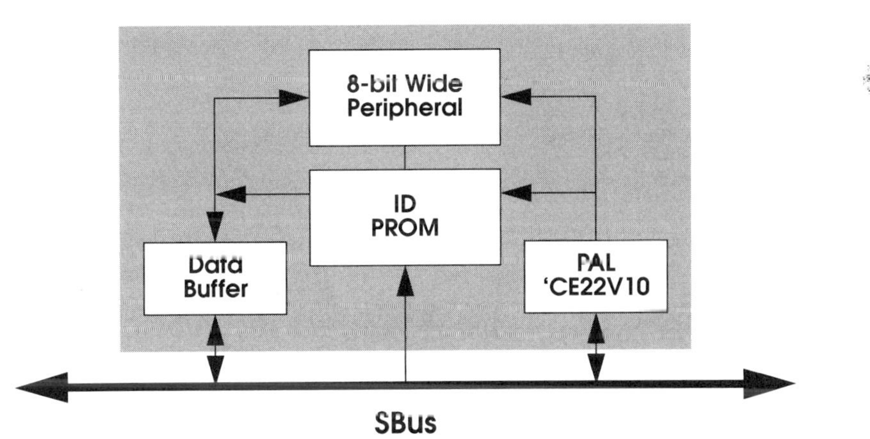

FIGURE 6.2. *Block Diagram of a Very Simple SBus Interface.*

This design has only four blocks. One is the requisite ID PROM, and the peripheral device is another. The data buffer is a 74FCT245 transceiver. Contrary to what might be expected, it is not needed to reduce loading on the data lines. With careful routing and component selection, the leakage and capacitance loads presented by the serial interface and the ID PROM together would not exceed those that the SBus allows. Instead, the buffer is needed because the time required to tri-state both parts exceeds the SBus' limit. The data buffer can be tri-stated much faster. There is no corresponding problem for the physical address lines; those are routed without buffering.

This design's PAL performs the control functions, and is a PALCE22V10-15 variety (CMOS, 22V10 architecture, 15 nanosecond input-to-output delay times). The functions it must perform include address decoding (qualifying it with **SEL*** and **AS***), data buffer enabling, write strobe generation, access timing, interrupt buffering, and **Ack*** generation. A more detailed drawing for this interface is shown in Figure 6.3.

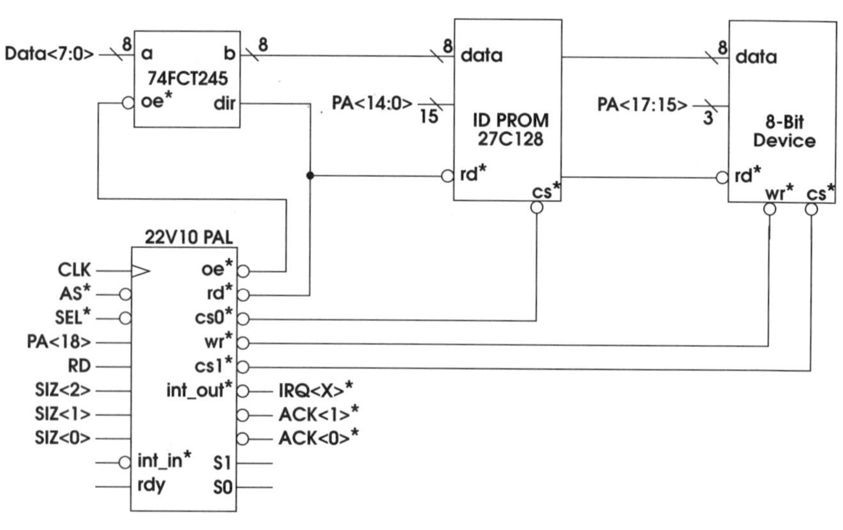

FIGURE 6.3. *Schematic of Simple Serial Interface.*

In this drawing, the interrupt and ready inputs to the PAL are not connected; the use of these signals is application dependent (they should be pulled high if not needed). The S0 and S1 pins should not be connected at all. They are state bits used internally. Notice that this interface only drives **Ack(1:0)***, because **Ack(2)*** need not be used in this case (see the related discussion starting on page 215). Notice also that the physical address bits are not shared. This reduces the load on these signals, at the expense of a larger address space. Careful routing and component selection can often eliminate the need to do this, or address buffers can be included instead. This circuit could be easily modified to support 16- or 32-bit devices, if needed.

FPGA devices are even more powerful than PALs when designing highly-integrated products. This is because a single FPGA can replace multiple PALs, as shown in Figure 6.4. This figure compares an AMD MACH 210 FPGA with the four PAL-CE22V10 PALs (the same type used in the previous discussion) that it can replace in most cases. The FPGA is more than equivalent to the four PAL devices. It contains more logic, and it also allows a larger number of interconnections between functional blocks (because signals don't have to go out through pins on one device and into pins on another). The package is smaller, there is lower total power consumption, and speeds are higher (again, because there are fewer chip crossings). The design of the I/O macro-cell is more sophisticated, too, allowing feedback from the I/O register *and* the pin. This is not possible in the PAL, which only allows feedback from the register (if a registered output) *or* from the pin (if a combinatorial output). This is a major improvement that reduces the logic required to implement internal registers that are both writable and readable.

Sophisticated SBus interfaces can be easily built with devices like the MACH 210, and that is why this type of device is used on the FCD10SBus fuzzy logic accelerator board. This FPGA performs the same functions which the PAL in the previous example did. Unlike that case, though, address decoding and access timing logic must be provided for seven different elements, not just two. This seven includes the two fuzzy co-processors, the two banks of knowledge base RAM, the ID PROM, and two control and status registers (one of which is also contained within the FPGA). Complicating matters, the port width and timing requirements of these elements vary. There are also two state machines (one per knowledge base) which control the buffers that multiplex the access paths to the RAMs. All this in a 44-pin rectangular package slightly more 0.5" (12.7 mm) on a side!

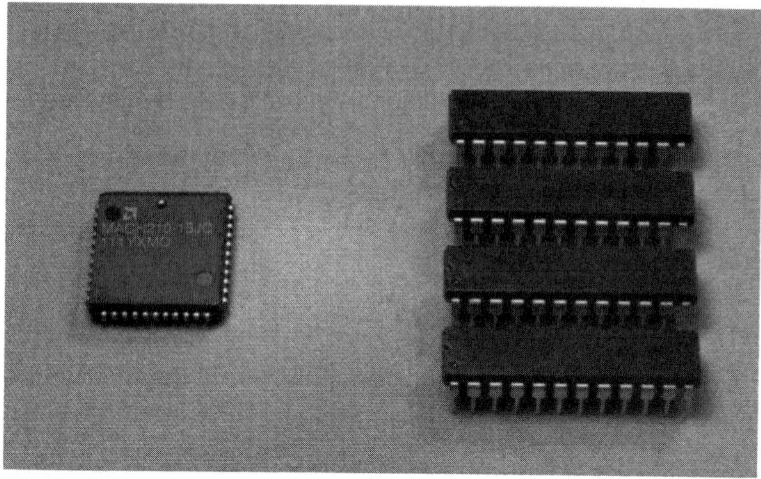

FIGURE 6.4. *Multiple PALs Can be Replaced by Single FPGA Devices.*

6.1.2 Surface-Mount Packages

Like CMOS, surface mount technology (SMT) is a key factor around which SBus was optimized. Surface mount is no longer a coming trend; it is a standard that is here now. Compared to a purely through-hole design, a surface mount device (SMD) based board can be designed with little additional cost or difficulty.

For example, consider again the FCD10SBus card, which makes heavy use of surface mount technology on both sides of the board. The assembly cost of this board is about $9.50 each in small quantities (20 - 30). There were some up front charges for tooling, solder-paste stencils, and "pick and place" programming, too. These totaled about $800, but are non-recurring charges and will not need to be paid for subsequent builds. Compare these costs now with those for this board's AT-bus based predecessor, which does not use surface-mount components. These boards cost $7.50 to assemble (there were no non-recurring charges for this board). This is $2.00 less per board, but much of that difference is because the AT-board has half the functionality, and hence fewer components.

Let's suppose now that sales volume projections for the product being designed total a modest 500 units. The total "buy-in" cost to use surface-mount technology probably averages no more than $3.00 or $4.00 per board, as it would in this example. That's hardly an insurmountable hurdle, and even that may be offset by savings in other areas. For example, perhaps a key component is available only in a surface-mount package (this is not unusual for some high-density, high pin-count integrated circuits). If the board were through-hole only, then a socket would be needed for a part like this. Such sockets are expensive, and even one might eliminate any possible cost savings of a through-hole only design.

Sockets are worth considering here for another reason, too. Some components need to be socketed. EPROMs typically are, for example, and other programmable devices such as PALs and FPGAs usually are, too. This is because the programming in these parts may need to be modified as bugs are fixed and features are altered. Such changes are especially likely early in a product's life cycle. Later, as the product matures and the changes grow less frequent, the sockets may no longer be needed. In fact, they might even become a liability because of the cost they add and the negative effects they can have on reliability. In cases like this it might be desirable to remove the sockets.

While many sockets for surface-mount components attach to the board with through-hole pins, others can be surface-mounted themselves. Some of these can even be mounted to the same trace pattern that the component they contain is designed for. This last is an advantage because it allows the socket to be eliminated at some future point without modifying the printed circuit board. If considering a socket like this, first determine whether the board assembler chosen has a reflow process which can reliably attach such a socket. Sometimes the plastic overhang interferes with the transfer of heat, and results in unreliable connections.

Surface-mount technologies can be used in most any product now with little added cost or difficulty. The benefits, though, can be substantial. The biggest benefit is that much higher levels of integration are possible; more logic and functionality can be packed into a smaller space. As this is one of the SBus' chief goals, it's no wonder that surface-mount technologies are such a good fit, pun intended.

There are several reasons why surface-mount packaging works so well in highly-integrated designs. One is that boards have two surfaces on which to mount components (see the related discussion which starts on page 207). Through-hole parts (unless oddly staggered) can only use one side or the other. Another reason is that SMD leads can be much more closely spaced than those of their through-hole counterparts. There is no pin that must fit through a mechanical hole in either the board or a socket. These holes ultimately limit the minimum pin-to-pin spacing because so many mechanical tolerances are involved. A drill bit can only be so small and can only be positioned with a certain accuracy. The hole must be plated with copper, which must have a certain minimum thickness, and the hole must have a pad, or annular ring, which has minimum dimensions as well. More importantly, auto-insertion equipment must be able to place a component with many pins, none of which is ever going to be bent exactly the way it should be. All these factors combine to set the minimum pin-to-pin spacing on through-hole components to the 0.1" (2.54 mm) which is currently standard.

Surface mount components don't have leads that must fit through holes, though; they need only attach to thin, usually linear "landing" pads or traces. Printed circuit technology is well suited to producing these even at very fine pitches, and so the component leads may consequently be placed at finer pitches, too. Spacings of 0.05" (1.27 mm) are commonplace, and even 0.025" (0.635mm) are gaining popularity for high pin-count packages. The obvious result is that a surface-mount package's pins will fit into a smaller space.

Surface-mount components are available in several different kinds of packages. Some are very similar to (but smaller than) DIP packages. Others are flat and rectangular. All can offer board area savings in most cases when used properly. Usually, it is possible to obtain either ceramic or plastic variants of any of these packages. The ceramic materials offer lower thermal resistances, but this is unlikely to be important in SBus environments unless Mil-Spec requirements are important. Plastic materials generally cost less, of course.

SOIC and SOJ Packages

Small Outline Integrated Circuits (SOICs) look very much like their bigger brethren in DIP packages, and the signal pin-out is usually the same. The packages are rectangular, with the leads spanning the two long edges. The leads are typically spaced on 0.05" (1.27 mm) centers, and are bent out away from package and down toward the board. This is called a "gull-wing" shape, because of its resemblance to a seagull's (inverted) wings. *Small Outline, J-*lead packages (SOJ) are virtually identical, except that the leads are curled down and under the package so that their shape is reminiscent of the letter "J."

Due to the finer lead-pitch and (often) narrower width of SOIC and SOJ packages, they typically require one-half to one fourth the board area of an equivalent part in a DIP package.

QSOP Packages

Quarter Size Outline Packages (QSOPs) are similar to SOICs, but about one-fourth the size. The leads on these packages are spaced on 0.025" (0.64 mm) centers. The sizes of equivalent 20-pin DIP, SOIC, and QSOP packages are shown in Figure 6.5.

FIGURE 6.5. *Size Comparison of Surface-Mount and Through-Hole Packages (equivalent DIP, SOIC, and QSOP packages are shown).*

LCC and PLCC Packages

Leadless Chip Carrier (LCC) packages are ceramic, and plug upside-down into special sockets which make connections to the conductive (usually copper) patterns on the package's surface. *Plastic Leadless Chip Carrier* (PLCC) packages differ not just because they are made of plastic, but because of the lead shape. PLCC leads are usually "J" shaped, as with SOJ packages. These do not require special sockets, and can be mounted directly to the board. They are also much more common than LCC packages. In either case, the packages are roughly rectangular, with connections made on all four sides of the chip.

The MACH210 devices used on the FCD10SBus card are 44-pin PLCC packages, which require less than 0.5 square inches (about three square centimeters) of board area.

QFP

Like LCCs, *Q*uad *F*lat-*P*ak (QFP) packages are rectangular, too. The leads are counted using a different convention, though, and the lead pitch is much finer, to accommodate higher lead counts. These packages often have 120 leads or more, with lead pitches as narrow as 0.025" (.635 mm) and even less. The leads may be either gull-wing shaped, or they may jut straight out from the package. The high lead counts of QFP packages make them well suited to ASICs and other pin-intensive applications.

6.1.3 High-Density Through-Hole Packages

Despite the many advantages that surface-mount packaging can provide, there will probably always be circumstances when through-hole packaging remains the best choice for some parts of the design. Perhaps mechanical strength is critical, for example. Components like sockets and connectors must often deal with mechanical forces and strains, and a through-hole part might be better able to absorb the punishment. Or perhaps a through-hole package is the only choice; that is still true of some components. And though the difference is dwindling, through-hole packaging is still generally less expensive, easier to handle, and easier to rework, too. For these and other reasons, improved through-hole packages are offered by many chip vendors. Some of these offer very high levels of integration which, in some cases, surpasses even that of the equivalent surface mount devices. Ultimately, this is perhaps the best reason why through-hole packages are still worth investigating closely.

Skinny DIPs

"Skinny" *D*ual *I*n-line *P*ackages (DIPs) were one of the earliest and simplest attempts to save board area. Many of the larger pin-count (typically 24 pins and above) DIP packages have the two rows of pins spaced 0.6" (15.24 mm) or more apart. Increasingly, though, IC vendors have opted for packages with pin rows spaced only 0.3" (7.62 mm) apart. Such packages occupy only half the board area of their wider counterparts, and there are few, if any, reasons not to use them if they are an option for the components you wish to use. In fact, the once "standard" (wider) packages are rapidly falling out of favor.

SIPs and SIMMs

Single In-line Packages (SIPs) seek to minimize the package's board area requirements by using vertical space instead. As the name implies, the package's pins form a single line along one edge of the package. The pins are in the same plane as the package, too, so that the component mounts seemingly on edge. SIP packages are often used for multiple resistors, capacitors or networked combinations of the two (R-C filters, Thevenin equivalent terminators, and other such combinations are commonplace). Some active components, such as serial NVRAMs and EEPROMs are available in SIP packages, too, but this has not caught on widely due to the relatively limited number of pins (usually SIP packages have no more than about 10 pins).

Single *In-line Memory Modules* (SIMMs) are an interesting variation on SIP packaging. As the name suggests, these are most often used for high-density memory applications. Each module is usually a hybrid; a small printed circuit board with several chips mounted and interconnected on it. Some modules have pins, but the vast majority are designed for use in special sockets. SIMMs offer very high packaging densities, and are now almost ubiquitous as a result. If considering SIMMs for use in an SBus application, care should be taken that the SIMMs used are not too tall. Also, because SIMMs are hybrids, some of the pins may present heavy electrical and capacitive loads. Traces on the SIMMs' PC board may also add to stub-lengths.

ZIPS

Zig-zag In-line Packages (ZIPs) are similar to SIP packages in many ways, except that the pins are not constrained to a single line; they are staggered in a zig-zag pattern. This allows more pins (20 or more are commonplace), which makes ZIP packages suitable for a wider variety of applications. It also enhances mechanical integrity, which may be important for some particularly tall packages. Finally, it improves routability; the staggered pins make it easier to route traces into and out of the component.

Many types of parts are now available in ZIP packages. These include memories, buffers, transceivers, and so on. Some IC vendors, such as Quality Semiconductor (Santa Clara, CA), offer ZIP packages across virtually their entire line of products.

The narrow footprints of ZIP packages allow them to be very densely placed on a PC board. Two of them will easily fit into the same space that a single equivalent DIP package would occupy. For these reasons both the FCD10SBus card and the SERFboard (described in Chapter 9) use ZIP packages for the address buffers and data transceivers that must be close to the SBus connector. An example is shown in Figure 6.6.

FIGURE 6.6.   *Close-up of Buffers in ZIP Packages.*

6.1.4 Other High-Density Packaging

For very sophisticated designs there are component attachment options which can result in very high levels of integration. These include the Tape-Automated-Bonding and Chip-On-Board technologies discussed in this section. These are cutting-edge technologies, and can be expensive. For certain designs, though, these options may be less expensive and require less time than would be necessary to design a large-scale ASIC. This is especially true for very specialized devices whose projected sales volumes are too low to justify an ASIC.

Tape-Automated-Bonding

Tape-Automated-Bonding, or TAB, technologies allow for very high densities because IC dies can be mounted without first packaging them in a bulky plastic package. Instead, a flexible "tape" or printed circuit board is used to mount the IC die and make the interconnection between it and the outside world. TAB techniques can increase the packaging density by a factor of five to ten times or more. In some cases, too, hybrids consisting of several IC die can be built on a single TAB module, further reducing real-estate requirements.

TAB techniques do have some disadvantages. For example, TAB is sensitive to mechanical flexing and twisting. This should not be a problem on most SBus cards, though, because they are small enough to be quite mechanically rigid. Some components are not yet available on tape, either, and not all printed-circuit board manufacturers can yet handle those that are. TAB is also a moderately expensive technology for the near future, although this can be at least partially offset for high pin count components, because of cost savings on the IC package that would otherwise be needed.

Chip-On-Board

Chip-On-Board, or COB, is just what you might expect. IC dies are mounted directly on the board (or sub-board). This can provide very high packaging densities.

Like TAB, COB is also sensitive to mechanical twisting and flexing, and can be moderately expensive. With COB there can be additional concerns because components may not be fully testable until mounted. Therefore this kind of technique is often best reserved for high-yield components.

6.1.5 Be Innovative with PCB Design

Often, not enough thought is given to the design of the product's printed circuit board. Some careful trade-offs and decisions can improve the routing efficiency of the board and maximize the number and kind of components it can contain.

Multi-Layer Boards

Printed circuit boards are made from one or more layers of a fiberglass material, which is usually clad in copper on both sides which is etched with the appropriate pattern. Multiple layers can be used in a single board, separated by an insulator and laminated together before being drilled and finished.

Very simple designs, or ones which are very cost sensitive, are usually built on two-layer boards. For boards like this, all traces (including those that distribute power) must be routed on either the top or bottom surface of the board. More complex designs usually require multi-layer boards. On these, one or more copper *planes* (surfaces) are often dedicated to power and ground distribution.

Multi-layer boards have far superior electrical characteristics than their 2-sided counterparts. The imbedded power and ground planes make for more constant impedances for signal traces on layers above and below. These planes also lower the power distribution impedance, and this simplifies bypassing, reduces noise and cross-talk, and improves noise margins. By far the biggest advantage, though, is that much denser boards can be routed because of the extra routing channels that multiple layers provide.

There are some disadvantages to multi-layer boards, of course. Multi-layer boards may take slightly longer to design and build, but especially in the latter case the difference is rarely significant. Most modern board layout tools can handle multi-layer designs, so there is probably little additional cost or effort associated with the CAD tools, too. The boards do cost more, of course, although not by as much as one might expect. Often, the cost of a printed circuit board depends more on the number of holes that must be drilled in the board afterwards, than in the number of layers the board has (the number of different hole sizes is a key factor, too, because each time the drill bit must be changed requires another step in the manufacturing process).

Usually SBus-sized printed circuit boards of up to 6 layers cost less than $100 each in small quantities, and $40 or less in volume. Using the 6-layer FCD10SBus card as an example, the cost per board in small quantities is $80.00, and the set-up charge totalled $620.00. The 2-layer AT-bus based predecessor costs $35 per board, and the setup charges totalled $465. The $155 difference in setup charges is almost trivial. The $45 difference in board cost is more important. Remember, though, that this distinction will diminish as volumes increase. Remember also that the SBus card packs about twice the functionality of the AT-bus card in about half the space. Capabilities like this may give your product an advantage over its competitors, or it may increase your product's appeal in other ways. This might make any slight increase in the printed-circuit board's cost a wise investment.

Mount Components on Both Sides of the PC Board

One of the best ways to highly integrate a design is to use both sides of the printed circuit board. This can nearly double the amount of board area that is available. The SBus card form-factor allows 4 mm (.157") on the solder-side for components. This is adequate for most active and passive surface-mount devices.

The FCD10SBus card uses both sides of the printed circuit board, as shown in Figure 6.7. Mounting some of the components on the solder side adds only about $1 to the assembly cost of each of these boards.

FIGURE 6.7. *Components Can be Mounted On Both Sides of the PC Board. In this case the bottom, or "solder" side.*

When considering mounting components on the solder side of an SBus card, make sure to also consider how these components will be probed during de-bugging. Access to components on that side of the board will normally be quite limited, but there are several ways that it can be accomplished. For example, the signals that need to be probed may also be available on pins of components mounted on the component side. Or perhaps there is a via (feed-through) which can be probed instead. Another possibility is to tack a wire to each signal that needs to be probed, and bring that wire around the board's edge. Or a right-angle extender board can be used, as shown later in Figure 8.3.

Consider Double-Tiered Boards

Another way to increase usable board area is to build two boards that can be stacked vertically as shown in Figure 6.8. This double tiered, or "sandwiched" approach has advantages beyond just increasing board area; it can also increase the modularity of the design. For example, assume connectors are used both for signal routing and for mechanical support. This allows the uppermost board to be removed or exchanged easily. At least one co-processor on the market already uses this as a feature that allows easy upgrades to higher-performance processors. This scheme might also be used to allow the end-user to mix and match I/O configurations, amounts of memory, and so on.

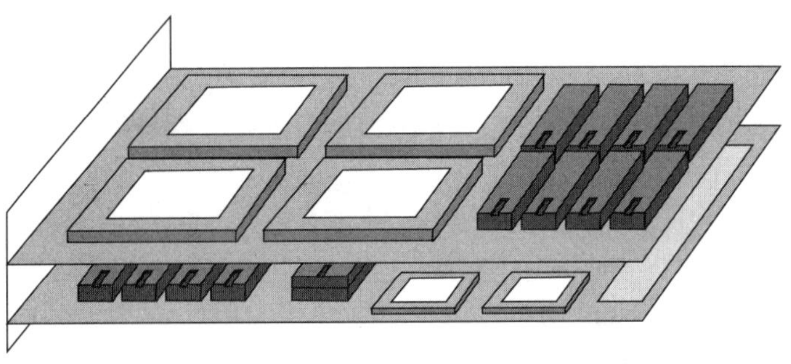

FIGURE 6.8. *Double-Tiered, or "Sandwich" Board Structure.*

Before deciding to design a double-tiered board, there are a few issues that should be considered. First of all, this may make it difficult to debug the card because access to components is heavily restricted. It is wise to develop some testing plan early in the design cycle, so that any circuit modifications or special components can be easily incorporated. Another issue is that this type of structure may be difficult to cool, and may block airflow to boards further "downstream" of the airflow.

Use the Space under Sockets

It is also possible to reclaim board area from underneath other components. This is especially useful when dealing with large components, such as EPROMs, or with socketed components. One example of how to do this is shown in Figure 6.9. Here, an EPROM is raised above the board using a tall socket. If the socket has a hollow space, that area may be used for other components. Socket strips (socket pins in a single-in-line configuration; two are used for each DIP package) work particularly well here because they are narrow, and leave a lot of space between rows.

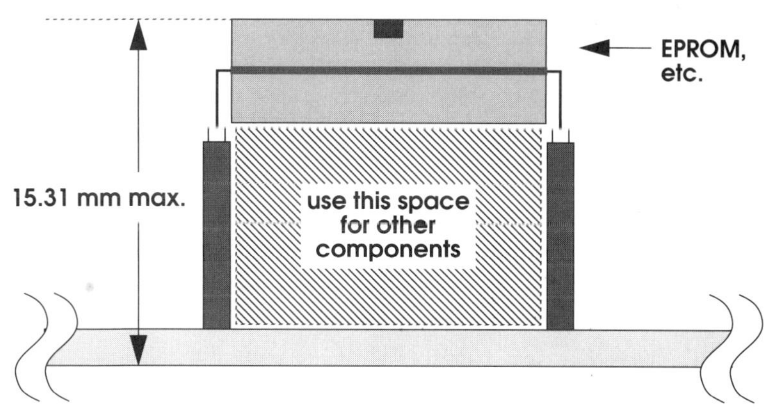

FIGURE 6.9. *Example of How to Use the Space Underneath a Socket (Front View).*

There may be 9 mm or more height available underneath a component mounted this way. Even when there is much less than that, there will often still be room for passive components, such as resistors and (bypass) capacitors. Most surface mount devices will work well, too.

If considering this option, there are a few caveats that need to be considered. First, access to any parts buried underneath others is restricted. This may complicate debugging. Also, cooling may be more difficult, too. First, the parts underneath may not get much airflow. Also, the stacked components may restrict airflow to other components "downstream."

Many manufacturers design special de-coupling capacitors made especially for use underneath the chip to be de-coupled.

These generally work well as long as the lead inductance is low enough. Also, there is a tendency for these capacitors to soften or melt when the board is wave- or reflow-soldered.

Maximize Board Routability with Careful Placement and Pin Assignment

Often, the first limit encountered when doing an SBus board layout isn't a lack of board area, but a lack of routing channels and especially of through-hole via sites. The number of connections necessary and the stub-length limits tend to concentrate components close to the SBus connector. High density packaging (especially through-hole technologies) can occupy a lot of grid locations in the area, leaving relatively few for the vias which connect signals between board layers. Often, not all of those remaining can be used for vias, anyway, because this would block many or all of the signal routing channels in the board. This can force signals to seek detours, which adds to the stub-length, may require extra board layers, and may also require even more vias! Sometimes the problem becomes so acute that the board designer finds that some of the board's area must be set aside just for routing purposes. This is wasteful, especially since the area around the connector where this is most likely tends to be particularly prime real estate.

Limiting the number of vias and routing channels needed will help eliminate these problems and allow best use of the board area available. Careful component placement and pin assignment can help to accomplish this. Place components with many SBus connections near the connector, and keep those with few or no SBus connections out of the way. Try also to avoid situations where large numbers of signals must "twist" or cross over each other, such as that shown in Figure 6.10. Here, pins 1-7 on one IC must be connected to the same pins on another IC. It is only possible to run a few traces (shown in black) before pads become trapped and the remaining connections (shaded) cannot be made without using vias to jump between layers. Data and address paths especially are susceptible to being blocked like this.

Two possible solutions to this problem are also shown in the figure. The first of these, in the center, is a simple placement change. The pin-to-pin connections are identical to the first case, except that now they can be easily routed on only a single layer. This illustrates the dramatic benefits that careful component placement can provide. In addition to sliding packages around, try rotating them. This can be especially effective with PLCC, PGA, and other packages which are generally square.

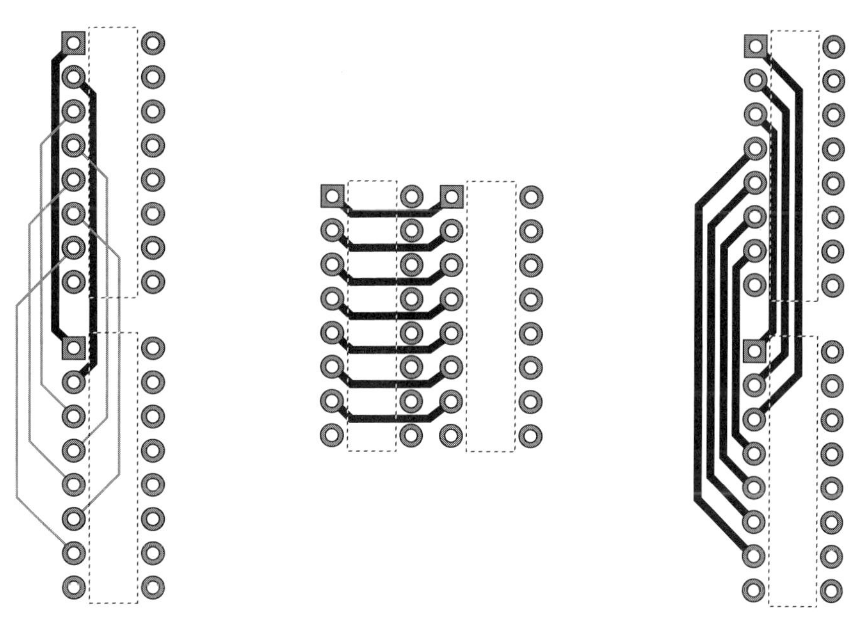

FIGURE 6.10. *A Routing Problem, a Better Placement, and a Better Pin Assignment.*

Sometimes, though, placement cannot be modified because of other factors. Even in these cases, for many components the assignment of signals to pins is flexible. Examples of such components include some PALs, registers, buffers, quad NAND or NOR gate packages, multiple resistor SIPs, and so on. In such cases it may be possible to modify the net-list in a way that eases routing constraints. This is shown in the last drawing in the figure. Here, pins 1-7 on one IC still connect to pins 1-7 on the other, but in a different order. In this case the placement is the same as that in the problem case, but the routing can now also be completed on only a single layer.

Whether trying to optimize placement, or pin-out, or both, it is useful to study the pin assignment of the SBus connector, diagrammed in Figure 6.11. Here, the various types of pins are coded so that signal groupings become readily apparent. It is interesting to note that the data and physical address lines form two separate, tightly-knit groups. The control signals are not tightly grouped; they are thinly spread across the connector, as are the power and ground connections. Neither the placement nor the pin-out of the SBus connector can be altered, of course, but it can serve as the "seed" from which the rest of the board design grows as it "crystallizes."

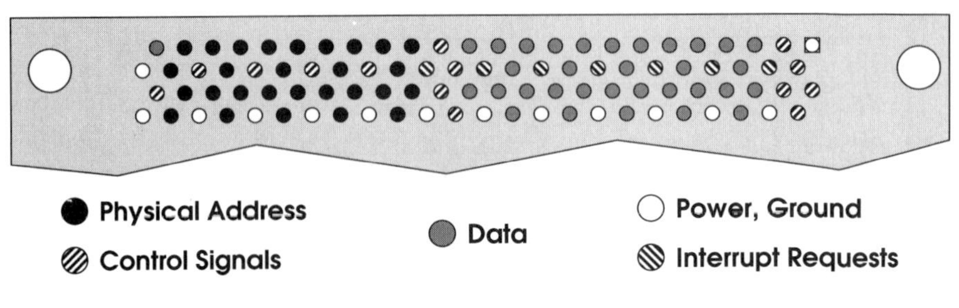

FIGURE 6.11. *Signal Locations on the SBus Connector.*

Sometimes the board designer runs into via site or routing channel shortages even after the component placement and pin assignment have been optimized as much as possible. At this point, "buried" vias may be one approach to consider. Normally, vias are plated holes that pass through every board layer, even though a connection is only necessary between two layers. A buried via is different because its plated hole connects only the layers absolutely necessary. Signal traces or other vias can use the same grid location, as long as they use different layers. This type of technology can greatly increase the routing density of a board. It is more expensive, but the board will probably route faster and with less frustration. Trace lengths will probably be shorter, too, and the board may hold more components or allow greater functionality.

Routing on SBus Hosts

SBus trace routing on the host motherboard can be difficult. The SBus connectors are usually placed edge to edge, and signals are bused in a way that virtually guarantees the kind of routing problem shown in Figure 6.10. Modifying the placement usually isn't a viable solution, because double-width SBus boards require a very specific relationship between connectors. Pins can't be re-assigned, either, for obvious reasons. Therefore SBus host routability can be a real problem with seemingly few solutions. Routing channels can be burned up very quickly. This may result in extra layers, and excessive trace lengths. In turn, this increases costs, propagation delays, and trace capacitance.

Adding significantly to the problem are the mounting brackets found on some SBus connectors (especially the surface mount variety). These require that relatively large holes be put in the board exactly where routing space is most needed. If building an SBus host, try to use connectors that do not require mounting hardware. The 96 pins in the connector provide quite a substantial holding force on their own once soldered (and the connectors have some specially "crinkled" pins to hold them in place until then).

Also, if components are packed right up to the SBus connectors this will reduce the amount of routing space available. This is one situation where it may be wise to dedicate some board area to the routing problem; leave some space for dedicated, un-obstructed routing channels.

In at least one recent case these precautions alone have saved two signals layers in the host's final board stack-up. The SBus trace lengths were also shortened, which improves its electrical characteristics.

Plating Equalizers

This discussion may seem out of place here, because it is not obviously related to reducing board area requirements. The relationship is indirect; this technique will help improve printed circuit board yields when fine geometries are used for trace widths and pads. Such fine geometries are necessary for some types of surface mount packages. More importantly, they can greatly increase routing densities, which in turn may lead to a more efficient use of board real estate.

After printed circuit boards are etched and laminated together, they are electro-chemically plated with copper (they are often "tinned" with a layer of solder, too). This plates-through the holes and vias on the board and leaves a uniform surface which will easily accept solder.

There can be problems if the traces and pads originally on the board are not uniformly distributed. Current densities in the plating tanks will not be uniform if this is the case, and plating thicknesses may vary accordingly. The result may be vias and holes that are inadequately plated, or plated so thickly that component leads cannot be easily inserted. Pads for SMT devices might also vary in height so much that reliable connections aren't possible.

In the past it was possible to adequately compensate for some of these problems because the technologies involved were less sensitive to plating thickness variations, and plating anodes could be adjusted to overcome some degree of imbalance. High manufacturing volumes and increasingly fine-pitched IC packages and board traces call for more advanced techniques, however.

The most obvious, of course, is to equalize the trace, pad, and via distribution. This can be done to some degree with careful board layout and routing. It can be optimized even further by including "plating equalizer" geometries where appropriate.

Plating equalizers are simply patterns of copper that can be used to fill voids in a routed PC board. Common patterns include simple cross-hatched grids, or arrays of round or square pads. Figure 6.12. shows a close-up of a section of PC board which includes plating equalizers. In this case the equalizers form an alternating array of small square pads, which fill areas that would otherwise be relatively blank.

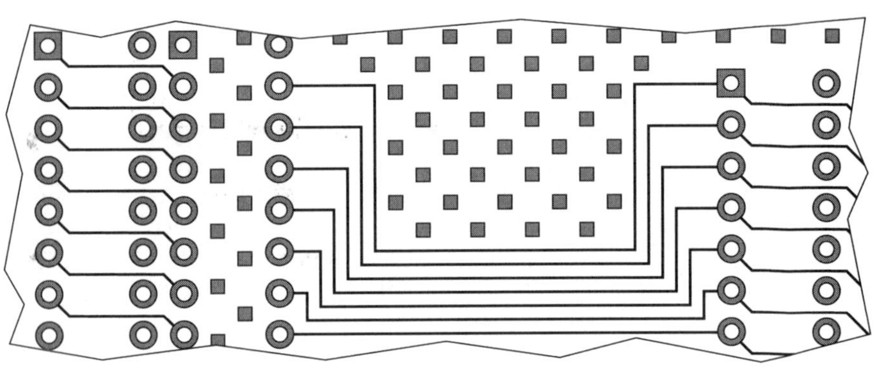

FIGURE 6.12. *Close-up of plating equalizer patterns on a PC board layout.*

6.1.6 Save Pins on ASICs and Other Parts Wherever Possible

At first it may be difficult to understand how saving pins on ASICs, PALs, buffers, and other parts can save board area, too. What difference will a few pins here and a few pins there make? It can make a lot of difference! For example, every single one of the 44 pins on the FPGA designed for the FCD10SBus card are used. If even one more pin were required then a bigger FPGA would have been necessary. The next size up in the family has 68 pins, is much more expensive, and requires over twice the board area! Alternatively, the design could have been re-partitioned, but this would probably have meant more parts. In turn, that also means more time, money, and board area.

Often, not all Ack* Signals Are Needed

Consider an SBus slave which is only a single byte wide. This slave is capable of servicing byte, half-word, and word transfers (in the latter two cases bus sizing will occur). This slave will use a byte acknowledgment (binary code 101) to accept these transfers, and an error acknowledgment (binary code 110) to reject all others.

Notice that during the byte acknowledgment only **Ack(1)*** is driven low, and during the error acknowledgment only **Ack(0)*** is driven low. The **Ack(2)*** signal is not driven low for either of these acknowledgments, though. Pull-up resistors normally hold all **Ack*** signals high when un-driven, and so this slave need not drive **Ack(2)*** at all!

Likewise, a word-wide slave need never drive **Ack(1)***. This slave will use a word acknowledgment (binary code 011) for transfers that it can perform, and an error acknowledgment (binary code 110) for all others.

Further simplification is possible if you can be certain that error acknowledgments will only be necessary under the rarest of circumstances. In that case, the error acknowledgment can be generated by simply ignoring the transfer and not providing any acknowledgment. That will be done by the SBus controller instead, after the time-out interval has expired. In this case the byte-wide slave need only drive **Ack(1)***, and the word-wide slave need only drive **Ack(2)***. This short-cut is not an appropriate way to generate error acknowledgments as a normal course of action, because the time-out interval is long and a substantial portion of the SBus' bandwidth could be lost as a result.

Map Out the ID PROM as Soon as the Driver Is Active

Another way to save pins on parts is to reduce the number of address bits that must be decoded. This is done by minimizing the amount of address space that the design uses. One way of doing this is to re-use address space, sharing it between two functions.

A good example of how this can be done involves the ID PROM. This is required on every SBus card, but is usually only read during slot probing. The contents are read, de-tokenized (decompressed), stored in, then executed out of RAM. Once the ID PROM is read, that address space isn't needed any longer. Something else, such as some buffers or another device, could be mapped instead.

The FPGA on the FCD10SBus includes this feature, although it is not used on the first revision of the board. A special I/O space can be mapped in place of the ID PROM. Fuzzy logic applications often include D/A or A/D converters and those would be accessed through this space. An internal register bit determines which space is active. It must be initialized so that the ID PROM space is always active after **RESET***. After that, the board's FCode or the driver can flip the bit's state and activate the I/O space. This feature saved one address bit on the FPGA.

Shadow register operations (described in section 3.4.5) can also be used to reduce the number of address bits necessary in any given design.

6.1.7 Save Parts Wherever Possible

One of the most obvious ways to save board space is to simply use as few components as possible. There are a variety of ways to reduce the number of parts used on a board.

Use 9 or 10 Bit Buffers Instead of 8 Bit Parts

Many common 8-bit wide buffer parts are also available in 9 or 10 bit-wide variants. This adds only a few pins to the package, but can save you entire parts.

This can be beneficial to your design even if you don't need the extra buffers for any given application. For example, if your design only needs 16 physical address bits, it might seem silly at first to use anything other than the two 8-bit wide 'FCT244 (or similar) buffers that would be traditional. If you did use 10-bit wide parts here, though, you might use the additional buffers for other signals, such as **RST***, **READ**, **AS***, **SEL***, etc.

Don't Use Special Drivers for the Interrupt Lines

The SBus' interrupt request lines must be driven by outputs which can only force the signals low, and can not force them high. Pull-up resistors on the bus are responsible for de-asserting the signals. This arrangement facilitates *wire-OR* connections (producing a logical OR function simply by connecting outputs together), and this allows the interrupt lines to be easily shared amongst several devices.

Outputs of this type are commonly called *open-collector* outputs, although *open-drain* is technically more correct for CMOS technologies. Unfortunately, such outputs are not easily found on off-the-shelf CMOS integrated circuits. Further, dedicating an entire part to driving what is usually just one or two interrupt lines is often not an attractive option. There are alternatives, though, that allow such outputs to be emulated in ways that require little, if any, board area.

For example, an open-collector or open-drain output can easily be emulated using a standard output and a Schottky Barrier diode, as shown in Figure 6.13. The diode is biased so that the gate can sink current, but cannot source it.

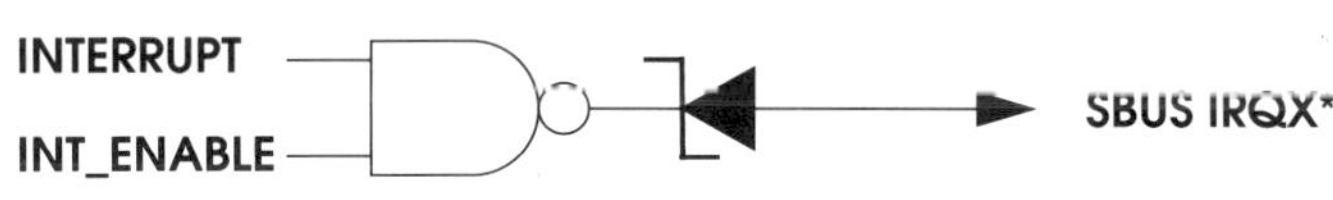

FIGURE 6.13. *Using a Schottky diode to emulate an open-collector output.*

Schottky diodes are used here for several reasons. They offer fast switching times, which is important in high-speed environments such as the SBus. They are also available with very low capacitance values; on the order of two picofarads. Capacitance in series does not add; the resulting total capacitance is always less than the smallest capacitance in the chain. Therefore the capacitive load seen by the bus is extremely small. It is for this very reason that newer bus driver technologies, such as Backplane Transceiver Logic (BTL) use series Schottky diodes to limit bus capacitance.

Most importantly, however, Schottky diodes also have low forward bias voltages, usually less than 0.4 Volts. This is important because the SBus' V_{IL} limit of 0.8 Volts must be met, and the V_{OL} limit of at most 0.4 Volts should be met. This last may seem difficult to accomplish, but it usually isn't. Most CMOS devices drive their outputs to very nearly 0 Volts in the low state, especially at the small current levels specified (4 mA) for the V_{OL} measurement.

Open-collector or open-drain outputs can also be emulated using tri-stateable drivers, which are already used in abundance in SBus designs. If it can be guaranteed that the driver is only enabled when it is to drive a low level, then its behavior will closely mimic that of an open collector driver, and it may be used to drive the SBus' interrupt request lines. When considering this, there are two cases to keep in mind. First, before the driver is enabled, the driver input must be set up with the correct logic level far enough in advance. Second, the driver input must be held valid long enough after the driver is disabled. The reason for this is to reduce the possibility of a momentary glitch as the driver is enabled.

One way to accomplish this is to simply tie the driver input high (or low if it is a non-inverting driver) and simply use the gate enable as the logic input. This structure is particularly well suited to many PAL and PLD families, and is shown in Figure 6.14.

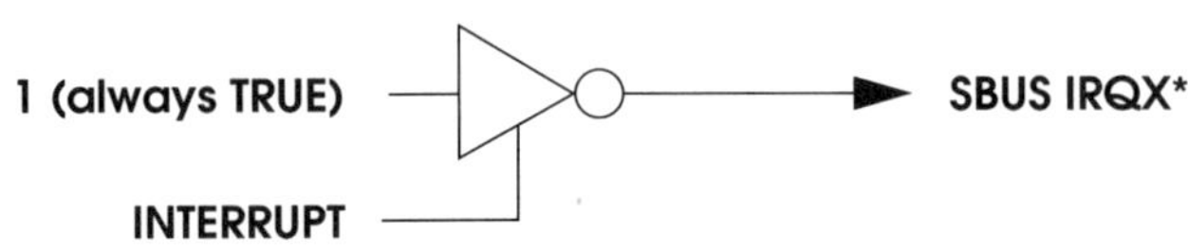

FIGURE 6.14. *Using a Tri-State Driver to Mimic an Open-Collector Output.*

This last mechanism works nicely with many PAL and FPGA output structures, and is used on the FCD10SBus card to drive the board's interrupt line. A fragment of the source code for the PAL which does this is paraphrased in Figure 6.15. This particular example uses the syntax of a PAL development tool called PALASM, offered by AMD. The syntax is very much like that used by other PAL development packages, as well. The pin is first defined as a low true combinatorial output. The input to this pin's buffer is tied high, and an interrupt is generated whenever the pin is enabled (in this case the enable is gated with **Reset***, to mask the interrupt until the board is properly initialized).

6.2 Minimizing Power Consumption

Another stumbling block often faced by those first considering an SBus design is the apparent lack of power available. Like the small form factors, though, the SBus' low power requirement is one of its most important advantages. Keeping the power consumption low simplifies cooling and reduces power supply (or battery) requirements.

```
PIN 23     /HIRQ                        ;combinatorial, low-true output
 .
 .
 .
           HIRQ=        1               ;buffer input is always high

; The interrupt is generated by enabling the output pin
        HIRQ.TRST =    HIRQ_ENB * /RST

; The interrupt can be generated when either co-processor interrupts
; or goes idle. Each source has a separate mask bit.
        HIRQ_ENB =     (S_INT_EN1 *INTR(1))+(S_INT_EN2 *INTR(2))
                       +(S_IDL_EN1 *IDLE(1))+(S_IDL_EN2 *IDLE(2))
```

FIGURE 6.15. *PALASM Code Fragment For Mimicking an OC driver.*

It is also important to remember that the power limits are based on the needs of CMOS technologies, which require only a small fraction of the power that would be used by the logic families that many designers have grown accustomed to. Consider the case of the SPARCstation 1+ workstation. The entire motherboard, including processor and system memory, consumes only slightly more power than that available to each and every SBus slot.

In the worst case, the FCD10SBus card uses about 8 watts of power. Typical dissipations are substantially less than that, and are a function of how heavily utilized the board is.

6.2.1 Take Advantage of the Technology

One of the best ways to minimize an SBus product's power consumption is to take advantage of the inherent low power characteristics of CMOS technology. The static power consumption of CMOS circuitry can be so low that it is often measured in microamperes!

Don't Switch Signals Unnecessarily

CMOS logic consumes most of its power when it is switching. One reason is that each CMOS gate incorporates two MOSFET transistors in its output stage; one to pull the output high and the other to pull it low (this is called a *totem-pole* configuration). These transistors are activated by voltage, not current, and their gate impedance is very high; once a voltage is applied to the transistor's gate, virtually no current (and hence power) is required to maintain it. Also, when the transistor is "off" its impedance is very high as well. The only steady-state currents that do flow are minute leakage currents. Normally only one of these totem-pole transistors is on at any given time, of course. While the output is changing state, however, there is sometimes a brief moment when both are at least partially conductive. This results in a temporary low-impedance path between VCC and ground, and this is where much of the circuit's power dissipation comes from (another is capacitive loading, discussed in the next section).

Power can be saved whenever signal transitions can be avoided or reduced in frequency. There are many ways to do this. For example, use Gray-code counters; a 4-bit binary counter makes almost twice as many transitions as a 4-bit Gray-code counter. Design state machines with as few state changes as possible, and assign the states in such a way that only one or two bits are modified whenever a state change does occur. Pick the lowest practical frequency for baud-rate clocks and the like. Isolate internal data and address lines (by disabling buffers) whenever an access isn't underway, so that every change or swing on the bus isn't repeated internally. Investigate and, if possible, use the features which allow the clocks of many components to be slowed down or stopped altogether when the device is idle. And keep the design as glitch-free as possible; each and every glitch has two signal level transitions.

Some of these suggestions may seem quite subtle, but when added up the effects can be substantial. Many data sheets include adjustment curves or specifications that allow the part's power consumption to be calculated as a function of clock (or input change) frequency. This information can be used to help quantify the effects of reduced switching frequencies.

Minimize Load Capacitance

Driving an AC signal into a capacitive load dissipates power. One way of looking at this is to consider a cup being rinsed over a sink. First the cup is filled up, then poured out, and then this process is repeated as often as necessary. This cup is somewhat analogous to the load capacitance. The amount of water (or power) required to rinse (drive) it is dependent both on the size of the cup (the capacitor), and on the rate at which it is filled and emptied (charged and discharged).

Mathematically, the capacitor is an impedance to AC signals which dissipates power much as a resistor would. The relationship is $P=2\pi CV^2F$, where P is the power dissipated, C is the capacitance, V is the signal's voltage swing, and F is the frequency (this formula assumes a periodic signal, but is generally applicable even for non-periodic signals if a good *average* frequency is used).

The power dissipated due to load capacitance is usually small at low to moderate frequencies, or inside integrated circuits (where capacitances are often very small). This can become predominant at higher frequencies, though, and at the higher capacitances found between chips and between boards.

Looking at the above equation, it is clear that reducing the power required to drive a capacitive load can be done in three ways. First, the frequencies need to be minimized. This can be done by limiting the switching rates, as discussed above, and by limiting the edge rates. Next, the voltage swing needs to be limited. This is an especially important factor because the power is proportional to its square. That is one reason that the SBus voltage levels are "TTL compatible" instead of purely CMOS levels; the reduced voltage swing results in power savings of 50% or more. Unfortunately, there is little the designer can do to minimize the voltage swing, except to rigidly favor technologies which do so themselves.

The final factor is one that the designer probably has the best control over. Capacitance should be eliminated wherever possible. This is another excellent argument for heavily integrating the design, because the capacitances inside a chip are much lower than those outside. There are a number of other ways to limit capacitance, of course. For example, the printed circuit board's cross section can be chosen so that it minimizes trace capacitance. Select and use parts that have a minimum pin capacitance, and avoid mounting them in sockets when possible (the socket pins add capacitance as well). Keep signal fan-out as small as possible, too, especially on high frequency signals.

Don't Let CMOS Inputs Float

It's rarely good design practice to leave an unused input unconnected. Whether the input pin is a "don't-care" on a larger function, or an input to an unused gate, it is usually better to tie the input inactive via a pull-up or pull-down resistor. Otherwise, the input is susceptible to switching accidentally due to capacitive or electromagnetic coupling with nearby signals. The input may also "float" to an invalid logic level.

These situations are especially likely with CMOS technologies because the input impedance is so high. Many CMOS inputs often won't tend to float into the high state, as do many TTL parts. As described in the preceding discussion, CMOS power dissipation is closely related to switching frequency, and unnecessary transitions waste power. If an input floats to a level near threshold, even a small amount of coupled noise can cause the gate to oscillate at a very high rate. If the gate's gain is low enough, too, an input level near threshold can turn on both transistors in the totem-pole output, effectively producing a near short-circuit between power and ground! In any event, power consumption is unnecessarily increased.

A simple pull-up or pull-down on unused inputs eliminates the problem. Here, CMOS technology's high input impedance is a major advantage, because one resistor can tie-off dozens of inputs.

The same arguments and the same solution apply to signals whose drivers can be tri-stated. Here, one resistor is needed per signal, but the high CMOS input impedances are still an advantage. A correspondingly high value for the pull-up or pull-down can be picked, and this will limit power dissipation in the resistor whenever it must be overdriven. The use of pull-up and pull-down resistors on the SBus itself is discussed in section 3.6.3.

There is also another way to prevent signals with multiple, tri-stateable drivers from floating. Design the circuit so that as one driver is turned off, another is always enabled! This scheme does not require any extra parts, and that is why it was used on part of the FCD10SBus card. The knowledge base RAMs are normally controlled by the fuzzy CPU, but the host must also be able to write and read them; this is the way they are tested and downloaded. The address and data lines of the RAM chips are driven either by the fuzzy CPU, or by the card's SBus interface. To do this, a multiplexor of sorts is built using tri-stateable buffers. A state machine was built into the SBus interface which alternately enables these buffers. When control of the knowledge base RAMs is switched, first one set of buffers is disabled, then after a brief pause the other set is enable. The pause is required to guarantee that the buffers are never both enabled at the same time. Its duration is only one SBus clock cycle, and is short enough that the lines' capacitance can hold them in their current state until the other buffer is enabled.

Use Power-Down States

Many components have *power-down* (or *standby*) modes which decrease the power they require during periods when they are relatively inactive. The power savings can be dramatic, especially in situations like static ram arrays, where only one bank is active at any one time, and all others can be powered-down.

Some components automatically power-down whenever they are not in use. Others require some conscious action, such as setting a bit in an internal register or activating a signal. One interesting technique is that used by some static rams. If the chip's select signal is de-asserted (raised to a level $\geq V_{IH}$), the part will automatically enter a partial power-down state. The chip will not enter the full power-down state unless the select signal is raised to within about 0.2V of the positive supply rail. This is easily accomplished using any CMOS logic family which drives its outputs from rail to rail. One good choice is the 74HCT family, for example, because its outputs have nearly the full 5 Volt swing, while its inputs are compatible with the "TTL" levels that are used on the SBus. An example of how this might work is shown in Figure 6.16. The pull-up resistor is used to help source the leakage currents on the signal and further guarantee an appropriately high voltage. In many cases it won't be necessary.

Be careful when using power-down modes. Sometimes this increases the access times of a component, or otherwise changes the way it operates. Make sure to take this into account. Also, current surges can result when powered-down parts are activated.

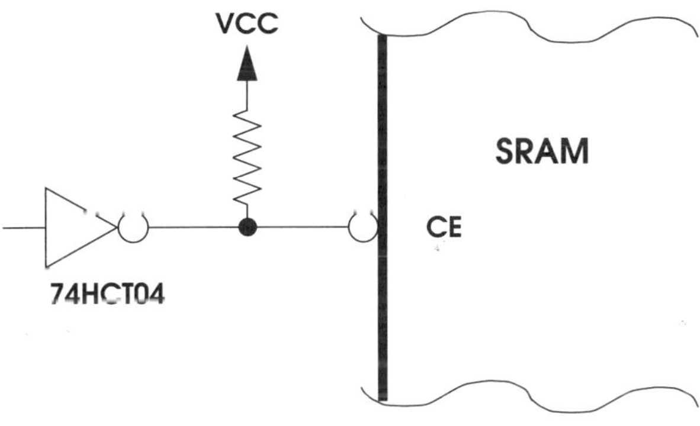

FIGURE 6.16. *74HCT logic being used to power down a static ram.*

Especially if a lot of them are activated simultaneously (in a bank of static ram, for example). Adequate bypassing is essential.

The set-up shown in this diagram is used on the FCD10SBus card to reduce power consumption by the card's knowledge base RAMs. Low power 32K x 8 CMOS RAMs are used, and each knowledge base is arranged in two banks of 16 bits each. The data books for the parts used show that for the cycle times required they consume approximately 100 mA each when active, but only 35 mA when in a power-down mode. Knowledge base accesses tend to be localized and relatively sequential, so usually only one bank will be accessed for reasonably long periods of time. This gives the other bank the opportunity to power-down, saving about 650 mW for each knowledge base and 1.3 Watts for the entire board!

6.3 Other Options

When all else fails, and your design still exceeds the SBus space or power budgets, there are still a few options available which can help solve the problems.

For example, a double-width form factor could be used. This is less desirable than a single-width card, because it won't fit in some machines which have too few unoccupied SBus slots. The advantage is that it provides twice the board area and twice the power budget.

Another option is to put only the SBus interface on the SBus card. The rest of the product is in its own external enclosure, and is connected via a cable to the host. This is already commonplace for additional disk, tape, or CD ROM drives. It is also used for products like serial-line multiplexors, often because the many serial line connectors require a lot of panel space. This external enclosure (sometimes called a "shoebox" because of it's general shape) can contain both logic and its power supply.

Sometimes, the logic will fit on an SBus card, and only the power budget is exceeded. In this case an external power supply is an option. For example, a wall transformer could be used, and connected to the board's logic through the backplate. When considering an option such as this there are a few points to consider. First, the external power supply must be separated from the internal, otherwise one or the other will find itself sourcing all of the current. Second, be careful not to dissipate too much extra power within the machine. This can cause problems both from a heat dissipation and a noise margin perspective; the host's cooling wasn't designed to handle a lot of heat from an outside source, and too much current flowing through the card's power planes and connector pins can cause excessive voltage drops and ground offsets. An ethernet interface provides a good example of how an external power supply might be a good solution. Some types of ethernet interfaces must supply 500 mA at 12 Volts to the external multiplexor/transceiver which is connected to the coax. This far exceeds the 30 mA which can be drawn from the SBus' 12 Volt supply. An external supply could provide it easily, though, and since the power is sent on to the transceiver and not dissipated in the host's enclosure, there are no cooling or noise-margin problems. Also, this supply is easily isolated from the host's, which is another ethernet requirement in some cases.

DC-DC converters can be used to convert one voltage into much needed current at another. These are most often used to bolster the 12 Volt supplies at the expense of the 5 Volt supply, but just the opposite is possible as well. They can also be used whenever a different voltage or an isolated supply is needed. Efficiencies up to about 80% are possible, and the cost and size is often very small, depending on the current requirements. Pre-packaged hybrids are commonplace and easy to use. It is also possible to build your own, using special integrated circuits and a small handful of passive components. This last approach allows the design to be tailored exactly to your needs, may allow for more efficient board area usage, and can save money if volumes are high enough.

Acknowledgments

The fuzzy logic board described in this chapter is produced by Togai Infralogic, Inc. #5 Vanderbilt, Suite A, Irvine CA, 92718. The product's model name is FCD10SBus. Special thanks to Doug Leo of Togai Infralogic, for all the effort and help he provided on this design effort.
Goldchip and SLIC are trademarks of Sun Microsystems, Inc.
LSI Logic is a trademark of LSI Logic Corporation.
PAL and PALASM are registered trademarks of Advanced Micro Devices, Inc.

References

Di Giacomo, J. ed., *Digital Bus Handbook*. McGraw Hill, 1990.
Ott, H. W. *Noise Reduction Techniques In Electronic Systems*. John Wiley and Sons, 1976.
Porry, D. "Multiple Technologies Shrink Future PCB Modules." *Printed Circuit Design*, February, 1991.
Glass, B. "Power Management." *BYTE*, September, 1991.
Sun Microsystems, Inc., *SBus Application Notes*, No.1, Revision A, July 1990.

Noise, ESD, and EMI Control

7

One very important aspect of any electrical design involves reducing its susceptibility to electrical noise related faults. Another element of the design involves reducing the circuit's ability to generate electrical noise and interference which might degrade other nearby circuitry. As most SBus products contain voltages and currents that are potentially hazardous, electrical safety is also a consideration, as is the prospect of damaging electrostatic charges (static electricity).

All these elements are closely related, and are the subject of this chapter. The discussion is broken up into two parts. The first part involves reducing the amount of noise that a circuit generates, and its sensitivity to noise from any source. The second part is concerned with isolating the circuit so that noise and interference can't get in or out. This last part is also useful for securing the circuit from safety and electrostatic concerns.

7.1 Reducing Noise Generation and Sensitivity

Electrical noise generated within a system can interfere with nearby electrical circuits, including the circuit which generated it! Noise is an additive phenomenon; while each individual source may be small, several of them may collectively wreak havoc. The best way to deal with electrical noise, of course, is to prevent it from being generated in the first place. If that is not possible, the next best alternative is to reduce the circuit's sensitivity to whatever noise is present.

Basically, electrical noise is composed of unwanted voltages or currents. A digital circuit's *noise margins* provide a measure of its sensitivity to electrical noise. These are the differences between a signal's voltage and the receiver's threshold (the approximate voltage at which it will change state). For example, SBus signals must be no higher than 0.8 volts in the low state, and no lower than 2.0 in the high state, measured at the receiver. Assuming that the receiver's threshold is about 1.4 volts, this means that there is a gap at least 0.6 volts wide between any logic level and the threshold. Any noise superimposed on the logic signal would have to exceed this noise margin before it might cause the receiver to switch inadvertently (noise margins are inexact measurements because receiver thresholds vary).

Electrical noise has several possible causes. One common cause can be signal reflections, which result in ringing, overshoots, and undershoots. Another frequent cause is the sudden shifts in a circuit's current demands as it switches. These generate voltage offsets in the power or ground rails due to inductive and resistive impedances. Noise on the power rails can also occur when it is conducted through an integrated circuit from one of its inputs or outputs. The same process can happen in reverse, as well.

The following subsections discuss means of reducing signal reflections (via signal terminations) and power-rail noise (with proper bypassing and ground-bounce suppression).

7.1.1 Proper Termination Techniques

Parallel terminations or Thevenin equivalent networks are used by some buses to guarantee un-driven line levels, and to "terminate" the transmission lines, thus reducing signal reflections. The SBus is not terminated. High value pullup resistors (or holding amplifiers) may be used to hold control signals de-asserted once driven there, but they are not effective at either pulling that line inactive or as transmission-line terminators.

Terminators are not used on SBus because they add capacitance and/or DC loading to the bus. This can greatly increase the amount of power required to drive the bus, especially at the high rates required. That is incompatible with SBus' original design goals.

Still, transmission-line effects are both real and significant. Excessive "overshoots" and "undershoots" may damage CMOS circuitry or cause it to behave erratically. Signal "ringing" may result in false clocking, or excessive settling times which will likely result in timing violations.

It is important to understand the transmission line principles involved, and the management techniques that are possible. In this way the card or host designer is better able to produce a design which is more robust and reliable.

Transmission Line Considerations

A complete discussion of transmission lines is far beyond the scope of this book. Only a few of the most basic concepts are presented here in the context of the SBus.

Transmission line effects are fundamentally related to the fact that electrical signals do not travel instantaneously. Ultimately they are limited by the speed of light, and practically (in wires or PC board traces) they are limited to some fraction of that.

Consider the circuit shown in Figure 7.1. Here, a voltage source applies a stepped pulse to a resistor. When the voltage pulse is applied, a current will flow into the line, through the resistor, and back through the return line. How much current, though? Ultimately, Ohm's law dictates that the steady state current will equal the step voltage V divided by the resistance R. If the resistor is close enough so that the propagation delay to it is effectively instantaneous, then the voltage step will appear across the resistor more or less unchanged, and the current will step directly to that expected. This is shown in Figure 7.2. This is often called a *lumped* load because the circuit behaves as if the loads were lumped nearby the source. Transmission line effects are negligible.

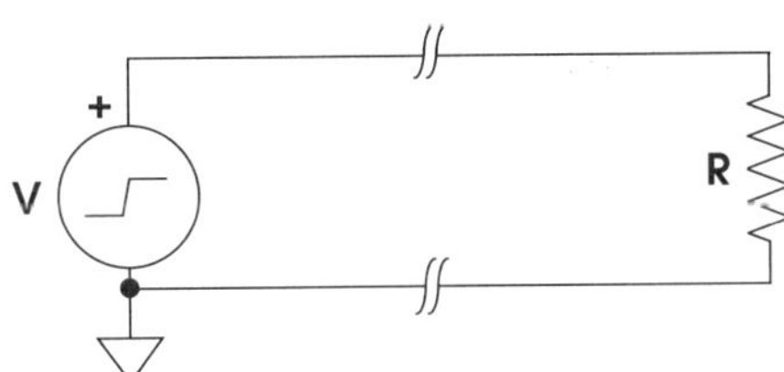

FIGURE 7.1. *Simple Transmission Line Model.*

The situation is very much different if the load resistor is far enough away. In that case, the initial current flow from the voltage source cannot depend on the load resistance, because the signal has not had a chance to get there yet. Instead, the initial current flow will depend mostly on the signal's line impedance. For the purposes of this discussion the line impedance can be considered a substitute load resistance which the source "sees" until the actual load resistance can be "found" by the traveling signal.

Line impedance (usually denoted by a 'Z' or 'Z_0') is a function of many variables. This includes distances and geometries of the layout, and the electrical properties of the materials involved (pc board, air, insulators, etc.). The electrical properties of any loads distributed across the line (leakage currents, capacitance, etc.) are also factors. For most standard printed circuit board technologies, impedances between 50 ohms (for heavily loaded signals) and 100 ohms (for very lightly loaded signals) can be expected.

If Resistor is Close Compared to Propagation Delay:

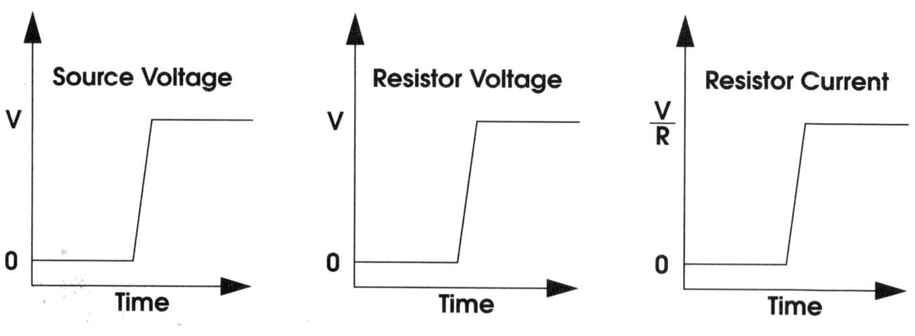

FIGURE 7.2. *Voltage and Current Waveforms for Circuit in FIGURE 7.1.*

Initial and final models for a transmission line are shown in Figure 7.3. As the voltage wave first starts propagating down the line at T=0, the value of the current wave associated with it is equal to V/Z. By the time everything reaches a steady state, though, the current will equal V/R.

The ideal situation occurs when Z=R. In this case the voltage and current waves launched by the source reach the load resistance after the finite propagation delay elapses. The voltage wave launched excites a current in the resistor that exactly matches the source's current, and everything immediately balances. From the source's perspective the steady state is reached at T=0, and the circuit is equivalent to an infinitely long transmission line. From the load's perspective, the steady state is reached immediately after one propagation delay time.

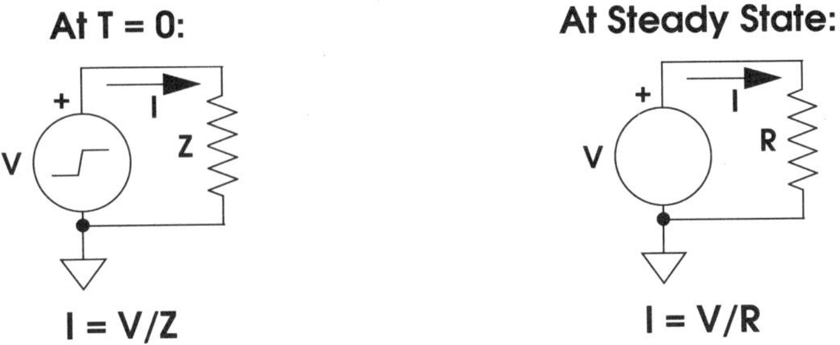

FIGURE 7.3. *Initial and Final Transmission Line Models.*

The situation is more complex if Z≠R. When the voltage wave reaches the load, the current created in the load resistance will not match the current wave launched from the source. This is a fundamentally incompatible situation because the currents into and out of any node must sum to zero. Physics provides a ready solution, however. The current imbalance can be erased by reflecting a voltage waveform back at the source. This waveform is superimposed on the voltage already present, and its magnitude is a function of the load resistance, the line impedance, and the source voltage. The relevant mathematics are shown in Figure 7.4. The constant $(R-Z)/(R+Z)$ is often referred to as the *reflection coefficient*.

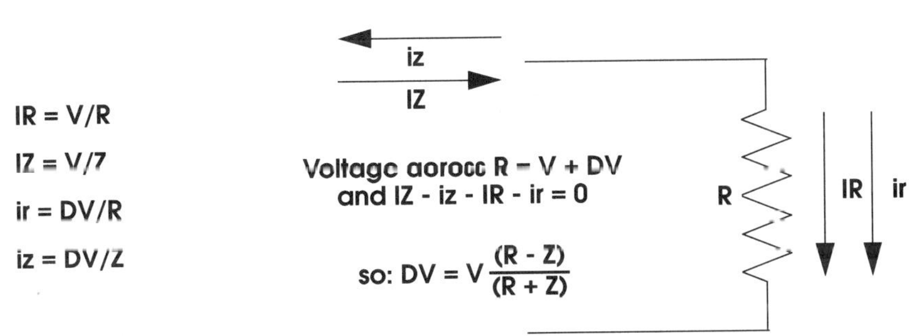

FIGURE 7.4. *Equations governing transmission line reflections.*

If R=Z, as in the first instance, the reflection coefficient reduces to 0. There is no reflection. If R is much smaller than Z then the reflection coefficient approaches -1, and the reflected wave nearly cancels the incident wave. If R is much larger than Z then the reflection coefficient approaches +1, and the reflected wave nearly doubles the incident wave!

An important point to keep in mind is that the reflected wave may cause yet another reflection when it returns to the source, and this process may iterate several times. Multiple reflections are the cause of the characteristic ringing (and overshoots and undershoots) which can be seen with an oscilloscope, as shown in Figure 7.5. Also, reflections don't just occur at the ends of a transmission line. Any change in impedance along the way can cause reflections, too. Connectors, IC pins, PC board vias, and bends or forks in traces are other possible sources of reflections. Even the scope probe used to look for reflections may cause them (or the reflections may exist only in the probe's coaxial cable).

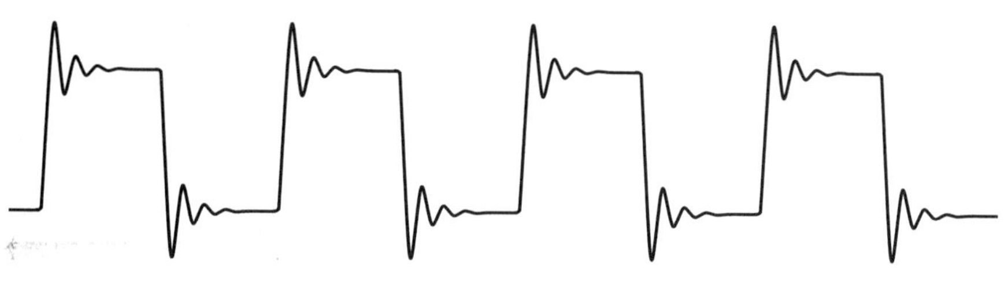

FIGURE 7.5. *Ringing caused by multiple transmission line reflections.*

Circuits in which transmission line effects are significant are usually called *distributed* circuits. There is no clear point that separates lumped and distributed circuits. Many circuits display characteristics of both, in fact.

One rule of thumb that is commonly used is that if the distance a signal travels in a circuit is some small fraction of the wavelength of the major frequencies, then the circuit is primarily lumped. Otherwise it is distributed. The fractions most commonly used are one-half or one-fourth.

For digital signals, the highest frequencies are components of the rise and fall edge rates. A useful simplification is that the period of the major frequency here is approximately equal to the edge rate. Propagation delays in printed circuit boards are approximately two nanoseconds/foot, so the approximate wavelength of a signal with 2 ns edges is about a foot. For this signal, then, trace lengths longer than about three to six inches will exhibit increasingly transmission line like tendencies.

Termination Techniques

It is important to limit reflections in transmission lines. Excessive ringing can significantly increase the time a signal takes to settle at a valid level. Reflections can pass through input threshold regions multiple times, causing false triggering and excessive power dissipation. Also, excessive voltage levels encountered might cause erratic operation or even damage. For example, an SBus driver might typically send a 4 volt signal into a 50 ohm transmission line. The load will most often be a CMOS input with a very high impedance (this assumption applies throughout this section). The result is that nearly an 8 volt signal impulse might present itself to the load! CMOS parts can easily latch-up or blow-up under such conditions. Internal registers might change state, and some PAL and PLD technologies have been known to partially "reprogram" themselves under even moderate undershoots.

One approach to managing signal reflections is to insure the circuit behaves as if it were lumped, not distributed. This is one reason that the SBus is a physically small bus (although no maximum length is specified). It is also the main reason that the SBus Specification calls for 5 ns minimum edge rates on all SBus signals.

Another method of dealing with reflections is to absorb them using clamp diodes. An example of how these are used is shown in Figure 7.6. The diodes are connected so that any input voltage more than a diode-drop outside of the voltage rails is clipped. The effectiveness of this technique varies considerably. Pin inductance and diode turn-on times are critical. Schottky diodes are often a good choice because they are fast and have low forward voltages. (Unfortunately they cannot be easily produced on CMOS substrates.) Despite what is occasionally lackluster performance, clamp diodes are becoming increasingly commonplace in many technology families. This is because they are compatible with other termination techniques, help prevent chip damage, and provide some level of ESD protection as well.

Ultimately it is better to prevent reflections in the first place than to absorb them once they are in the circuit. Resistive terminators are a traditional way of doing this. One method is to approximate the ideal case in which the load impedance matches the line

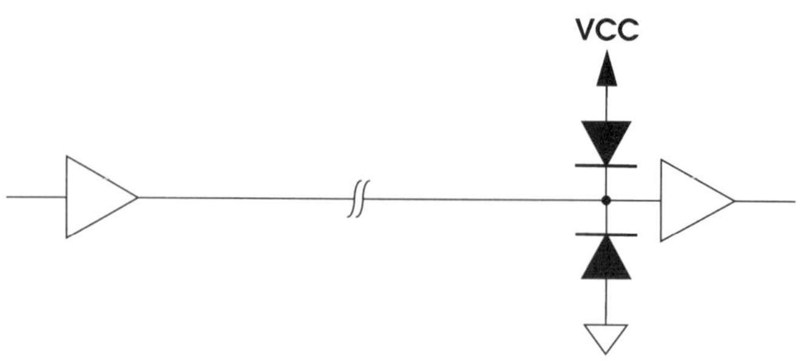

FIGURE 7.6. *Clamp-diodes used to dampen ringing and protect the receiver's input.*

impedance. This can be done with a single resistor to ground, VCC (+5V), or some other fixed voltage source (all of these cases are equivalent where termination is concerned because superposition applies and the mathematics work out regardless of the line's initial voltage). Alternately, a Thevenin equivalent network composed of two resistors in parallel can be used, as shown in Figure 7.7.

The resistors in the parallel network are chosen so that the parallel resistance is equal to the line's impedance, and so that the Thevenin voltage is at the desired level. This is often a valid logic 'high' voltage, which allows the resistors to perform both termination and pullup functions.

The waveforms associated with parallel termination mechanisms are also shown in Figure 7.7. As expected, the voltage waveform from the source travels to load B in one propagation delay time, passing load A halfway along. If impedances are well matched, there are no reflections.

One major drawback of such parallel terminations is the amount of power they consume. High-powered drivers are required to provide the large amounts of DC current necessary. The power level required can exceed that of SBus drivers by up to 100 times or more!

A less power-hungry method of terminating the line at the load is shown in Figure 7.8. This circuit uses a series R-C high pass filter to perform the termination. The values of the resistor and capacitor are chosen so that the filter's impedance equals the line's impedance at the primary frequency of interest. At or near DC,

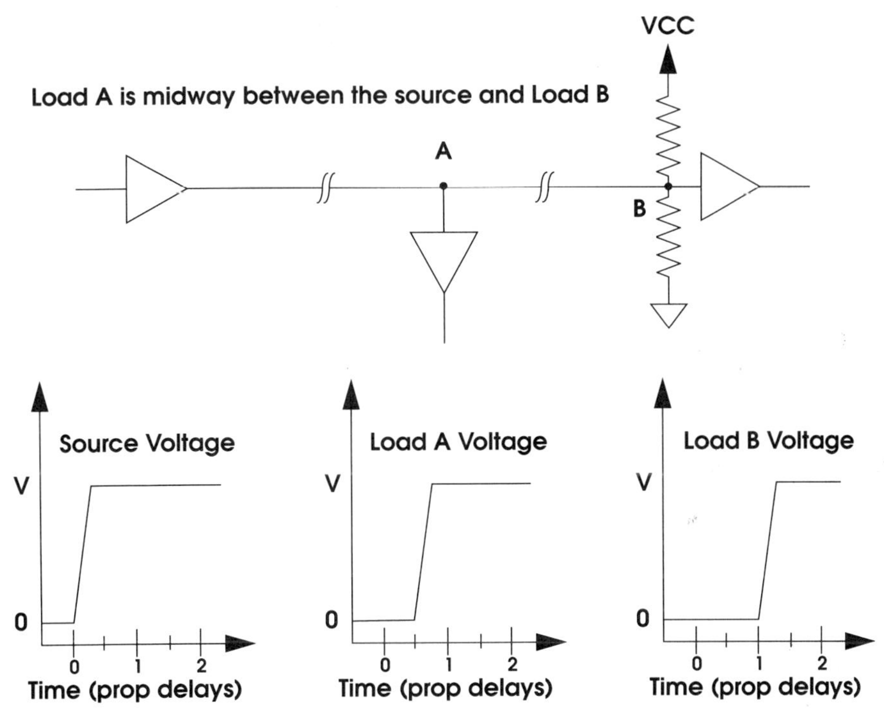

FIGURE 7.7. *Parallel termination of a transmission line.*

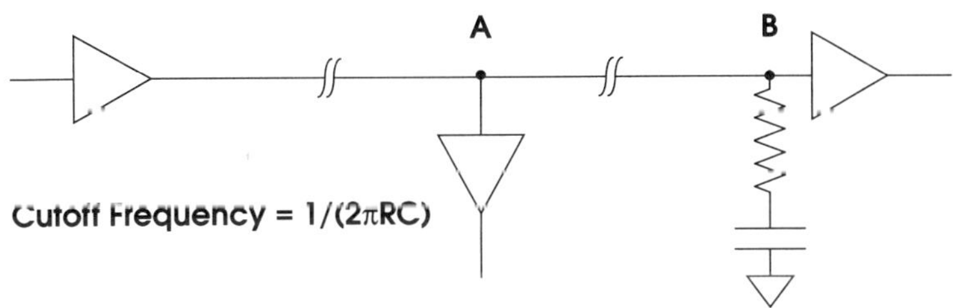

FIGURE 7.8. *AC termination scheme.*

though, the impedance climbs rapidly, reducing the amount of DC current the driver must source or sink.

There are some disadvantages to the AC termination scheme. The most important is that it is frequency dependent. If the cutoff frequency of the RC filter is too low or too high, the terminator won't be effective. Also, if the cutoff frequency is too high, this won't save much power, because the capacitor's voltage will reach roughly half the signal swing and stay there. If the signal being terminated is roughly periodic, though, this can work well. Make sure to select a frequency from the edge rate (not from the period of the signal itself) because these are the higher frequency components.

Yet another mechanism for terminating a transmission line is shown in Figure 7.9. This is usually referred to as series termination, because a resistor is placed in series between the driver and the transmission line. The resistor and transmission line then effectively form a resistive voltage divider. If the resistor is chosen so that its value is equal to the transmission line's impedance, then the voltage wave launched is initially equal to one-half of the driver's voltage swing. When this wave reaches the far end of the line, the load it encounters has an impedance much larger than that of the line. The result is a reflection coefficient of nearly +1, which in this case is beneficial. The reflected wave will nearly double the line's voltage as it passes again, bringing it near the original level the driver sourced! When this reflected wave reaches the driver again, the steady-state will have been achieved and no more reflections will occur.

The voltage waveforms at various parts of the circuit are also shown in Figure 7.9. For the driver and the final load, things are much the same as in the parallel and AC termination cases. Things are very different for loads located between these points, however. These loads see a distinct stair-step as the incident wave first passes bringing the voltage up halfway, then the reflected wave passes again and finishes the job. The timing of this stair-step varies along the line. Loads near the end will see almost none. Loads near the resistor will see a stair-step that lasts nearly one complete round trip (two propagation delay times).

Series termination does not suffer the power or frequency dependent drawbacks that the parallel and AC alternatives do. Because the signal must travel out and then back, however, it can take twice as long for loads close to the driver to see a valid signal (loads near the end will not see any appreciable extra delay). Also, the stair-step can be a problem because it will usually fall in or near the thresholds of the receivers. For this reason it should not be used for edge sensitive signals such as clocks, unless there are only a very few loads clustered near the signal's end-point.

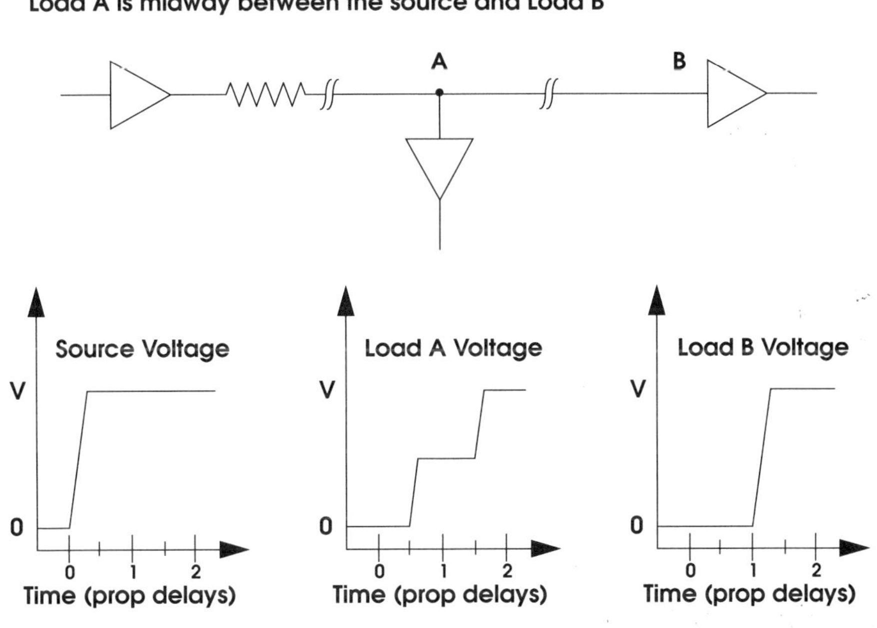

FIGURE 7.9. *Series termination of a transmission line.*

One other feature of series terminators can be either positive or negative, depending on the application. The line's capacitance (both stray and that of the receivers) will combine with the series resistor to form a low-pass R-C filter, which will slow the signal's edge rates down. If the impact this has on signal timing are not overly significant, then it can be a big advantage because it makes the bus act more lumped and less distributed.

Given all these properties, series terminators can be a good solution to transmission line management on the SBus. An idealized bus structure using series terminators is shown in Figure 7.10. Transceivers are connected to the bus; two at the ends and one in between. The input impedance of the transceivers is much greater than the series resistance added by the terminators, so receiver function is not greatly affected.

One interesting observation is that drivers connected between the ends are connected to two transmission lines; one to either side. The impedance of these lines adds in parallel, and so the value of the series terminator required is half that of the line's overall impedance. If building a host, you have control over signal routing and line impedance, and can easily pick the right resistor value.

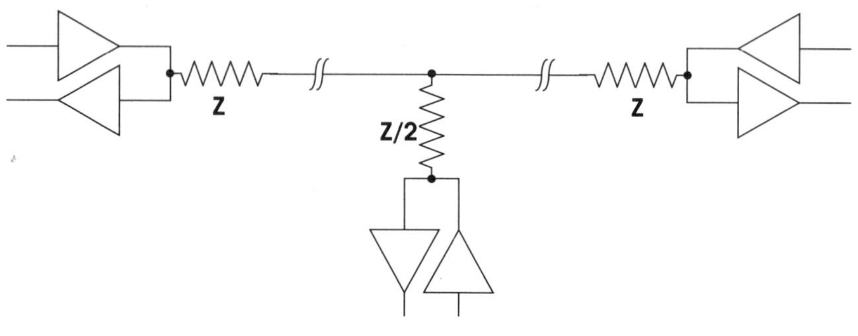

FIGURE 7.10. *Bus structure with AC terminators.*

If you are building an add-in card, you do not know whether you will be at the end of the line, except for the radial SBus signals (such as **BR***). You cannot be sure what the line impedance will be, either. In this case it is reasonable to assume that your card will be placed in the middle of the line, and that the line impedance is 50 ohms. This yields a terminator value of 25 ohms. Even if these assumptions turn out to be wrong, the results are much better than if no termination were used. For example, assume the card really is at the end of line, and the line impedance is as high as 75 ohms. The first reflection will still be reduced by at least 25%. It is no coincidence that there are an increasing number of parts becoming available that have 25 ohm internal series terminating resistors (and clamp-diodes)!

Series terminations on drivers, and clamp-diodes on receivers can do much to improve signal quality on the SBus. Designers of both hosts and add-ins are encouraged to investigate these alternatives where possible.

7.1.2 Proper Bypass Capacitor Selection

Most designers are aware of the need to include "bypass" capacitors in their designs. These may also be called filter or "de-coupling" capacitors. Standard rules of thumb such as, "use one 0.1 (or 0.01) μF (microfarads) capacitor per IC (or every other IC)," have become ingrained and are almost automatic. Bypass-related problems have been rare, and as a result few engineers have had to consider the problem beyond this level. The SBus uses modern technologies with very fast edge-rates, though, which are not as forgiving as past technologies have been. It is important for an SBus designer to be aware of the issues involved and make an informed decision about the kind of bypass strategy to use.

When a digital logic circuit switches, the amount of current it draws from the power supply usually increases or decreases. When this happens, the supply voltage usually also changes in some way because the supply impedances and response times are finite.

The result can be a large amount of electrical noise induced in the power supply distribution network. This noise is undesirable because it can interfere with nearby equipment (via EMI). It can also disrupt local signals by coupling into them and detracting from their noise margins. If severe enough, this noise can even cause the local voltage to go out of spec, resulting in the improper operation of the logic circuit itself.

These problems can be reduced or eliminated by distributing bypass capacitors throughout the design. These capacitors are connected between the supply rails and ground, and work by lowering the AC impedance of the supply distribution network. Distributed bypass capacitors work like low-pass filters, and they store small amounts of charge locally, where it is more immediately accessible.

To think of this another way, consider the simplified circuits shown in Figure 7.11. Both schematics show a logic circuit at some distance from the power supply, with a finite impedance **R** in the power distribution path. This impedance is the sum of the power supply impedances, the actual resistances within the power wires or planes, and any inductive effects in the AC cases. For simplicity, it also represents the Thevenin equivalent of impedances in both the supply and the ground systems.

Suppose that as the logic circuit switches, its current demand increases or decreases. If there is no local bypass capacitor, the voltage seen at the load will either increase or decrease as the voltage induced across the distribution impedance changes with the varying current. A regulated power supply may be able to compensate for some of this, but its response times are usually far too slow to be of much use. Any filter's response time must be less than the period of the signal or noise that is being filtered. For example, assume we want to filter noise whose highest frequency component is 100 MHz. The minimum period of this noise is then 10 ns, and our filter must respond faster than that if it is going to be of much use.

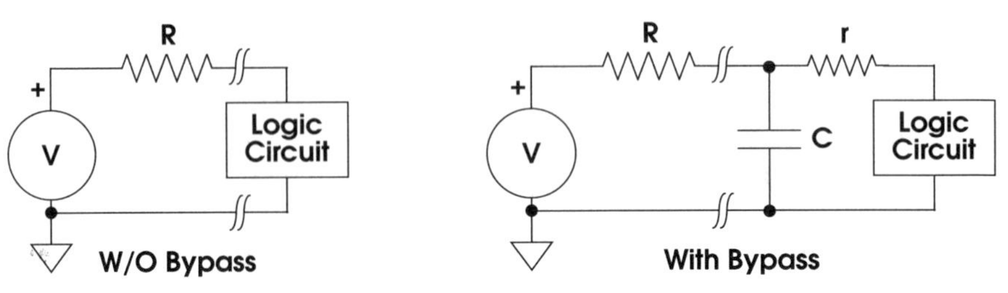

FIGURE 7.11. *Power Distribution Models.*

This is where the bypass capacitor can help. The capacitor stores charge locally. If the load momentarily increases its current demand, the excess charge can be *supplied* by the capacitor. If the load momentarily decreases its current demand, the excess charge can be *absorbed* by the capacitor. Overall, bypass capacitors average the load seen by the supply, reducing the frequency response of the overall distribution network. This is beneficial because it reduces inductive impedances, and it allows the power supply adequate time to be effective.

A number of criteria must be met for any bypass strategy to be effective:

- The bypass capacitors must be physically very close to the loads. Lead lengths must also be minimized. These are critical because they minimize the resistance and inductance between the capacitor and the load. Surface mount ("chip") capacitors can work very well.

- When specifying a bypass capacitor it is important to choose the *type* of capacitor, and not just its value. No capacitor is perfect. Each also contains stray resistances and inductances as shown in Figure 7.12. Above certain frequencies it is entirely possible for the capacitor to behave much more like an inductor!

A capacitor's specified *self-resonant frequency* is the point at which this happens, and for any given capacitor this is a function of the value, construction technique, dielectric material, and lead length. Mathematically, the self-resonant frequency = $1/(2\pi\sqrt{LC})$, where L is the sum of all lead and stray inductances, and C is the sum of the component capacitance and all other lead and stray capacitances (leakage and series resistances are assumed negligible).

- The value of the capacitor must be large enough so that the variations in current expected do not greatly change the capacitor's voltage during the time it takes the power distribution network to respond. This basically means that there is a minimum frequency below which the bypass capacitor is no longer effective (because it no longer presents a low-impedance path to ground). Interestingly, this is also partially a function of the capacitor's self-resonant frequency. A common rule of thumb is that this minimum frequency is one-third to one-fourth the resonant frequency.

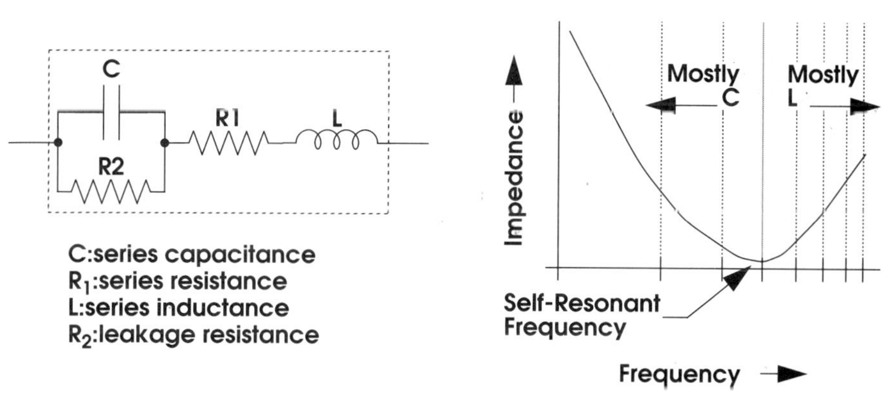

FIGURE 7.12. *Capacitor Equivalent Circuit and Typical Frequency Response.*

- Small IC packages can be bypassed with a single capacitor. Large packages (and especially large ASICs in pin-grid-array or leaded-chip-carrier packages) may require multiple capacitors.

In all cases it is preferable to place the bypass capacitor near the VCC pin(s).

As you might expect, these requirements often conflict. Any capacitor whose value is large enough is often physically large as well. This makes it difficult to place close enough to the components that need bypassing. Also, construction techniques and dielectrics that lead to large capacitance values tend also to have low resonant frequencies, and so are not effective for logic with fast edge-rates.

A common solution to this dilemma is to divide the task and include both small "fast" capacitors and bulk "slow" capacitors. The small value capacitors are placed right at the chips they are to bypass, as close as possible and with leads as short as possible. These capacitors will be responsible for filtering the highest frequency components of the noise. Larger value capacitors can then be used to filter the remainder of the noise. Their placement is less critical and fewer of them are needed. In very fast circuits sometimes the problem needs to be divided even further, into "fast", "moderate," and "slow" capacitors.

By now the basic concepts involved have been discussed, and it is time to discuss how this relates to your design, and to the SBus.

The SBus is optimized around CMOS technologies with very fast edge-rates. Transition times of 2 ns or less are easily achieved, and signals such as these can have high-frequency components of 1 GHz or greater! Fortunately, you do not need to worry about bypassing frequencies this high because stray capacitances in the IC's die, package, and the printed circuit board itself will do that for you. 100 MHz is a more reasonable target for the highest frequency that must be bypassed. Low-loss ceramic capacitors in the 100 to 1000 picofarad range are capable of this *if the leads are kept very short and they are close to the loads.* This effectively means that one of these will be needed for each potential source of high-frequency noise.

If this first line of defense is chosen to work at around 100 MHz, then it will probably not be very effective below about 25 MHz. These frequencies are still too high for most bulk capacitors, so more small-value ceramics are needed. In this case 0.01 to 0.1 microfarad capacitors would work well, and their placement is less critical. How many are needed? The sum of these capacitors' values should at least equal the sum of the values of the smaller capacitors. More typically a 5x or a 10x factor is used. This sounds drastic, but one such capacitor for every 10 or more of the smaller ones is usually ample. In practice, this means that a small handful are sprinkled around the board.

This second group of capacitors will filter effectively down to 6 MHz or so. This is still too high for most bulk capacitors, so yet another mid-value group of ceramic capacitors is needed; something in the 0.1 to 0.22 µF range. Only two to four of these are probably needed, spread around the board.

These capacitors will filter down into the range of 1 MHz or so. At these frequencies there is much less concern about electrical noise. This is because in an environment such as the SBus there are few sources that can generate noise frequencies this low! Also, what noise is generated is less likely to cause problems or be an EMI issue (which primarily concerns higher frequencies).

At this point, then, it's usually appropriated to put in the bulk capacitors. Mylar, paper, or tantalum capacitors are best suited to this purpose. A general rule-of-thumb for bypassing is that several small capacitors are better than a few large ones, and this is especially true for bulk capacitors. Values of about 5 and no more than 10 µF are appropriate. Capacitors that are physically small and have short lead lengths are best. The best place to put them is in the corners of the board, where they are out of the way and where they are evenly distributed with short average distances to all points on the board.

The result is a multi-tiered approach with at least four different types of capacitors. One example of a possible solution is outlined below:

- All active parts clocked at frequencies above 20 MHz and all parts known to have high edge-rates (especially buffers) are bypassed with a 680 pF capacitor. The capacitor should be a low-loss ceramic type in either a radial or surface-mount package. The capacitor should be placed as near as possible to the part's VCC pin. If the part has more than one VCC pin, then there should be one such capacitor for each.

- All other active parts should be bypassed with a 0.033 µF capacitor. The capacitor should be a low loss ceramic type in either a radial or surface-mount package. Additional capacitors are needed near clusters of the 680 pF capacitors. Minimally, there should be 4-5 of these capacitors distributed across the board.

- At least 2 4 0.22 µF capacitors should be distributed around the board. The capacitors should be a low-loss ceramic type in either a radial or surface-mount packages.

- There should be one 5 µF capacitor at or near each corner of the board. The capacitors should be electrolytic, of the mylar, paper, or tantalum type in either radial or surface-mount packages. If the design contains banks of memory, or other types of circuity whose current demands can change suddenly and dramatically, then more of these large bulk capacitors should be distributed nearby.

The reader is strongly urged not to accept this as a replacement for any rule of thumb which they have used in the past. This is only one possible solution, and may prove either over- or under-designed for any one particular application. By understanding the concepts involved, though, an adequate solution can readily be found.

7.1.3 Ground Bounce Suppression

Ground bounce is a very particular form of power-rail noise, of special importance with CMOS devices. The effect is most prominent on the internal ground level reference of an integrated circuit, which may "bounce" dramatically under certain circumstances as noise voltages are imposed on it. This can be a serious problem, because there is often a low impedance path from the ground reference to any output in the low state. This would allow the ground-bounce to pass out through a chip and undermine the noise margins of those signals. The integrated circuit's thresholds are also based partly on the ground level reference. If one shifts then the other does, too, and this can undermine the noise margins of inputs to this chip.

Like other power-rail noise, ground bounce is the result of sudden changes in current demands, and finite (mostly inductive) impedances in the power distribution. Proper bypassing will not help in this case, though, because the critical impedances are in the chip's internal lead-frame. Generally, at the root of the problem are low-impedance outputs which drive primarily capacitive loads at very high edge-rates. Unfortunately, this is typical of CMOS and SBus environments. In this situation, the driver must sink or source large currents for the short period of time required to charge the capacitance. The situation is particularly critical when a majority of a chip's outputs change together and in the same direction. The lead frame inductance converts this sudden current pulse into a voltage pulse, superimposing it onto the normal ground reference level.

One way to limit ground bounce problems is to choose components packaged in such a way that lead-frame inductance is minimized. Many manufacturers are sensitive to ground bounce problems and have optimized component packages to help eliminate it. In general, devices will have improved characteristics if the packages are physically smaller, have multiple ground pins, or have ground pins closer to the device's center than to its corners.

Ground bounce problems can also be avoided by limiting the magnitude and frequency components of the unwanted current pulses, and hence the effects of the lead-frame impedance. This can be done in several ways. For example, the designer should restrict the number of signals that can switch simultaneously, and in the same direction. This is best done by careful logic partitioning, state assignments, and signal polarity selection. Slew-rates (the minimum rise and fall times) should also be limited and output impedances increased. The series terminations discussed in Section 7.1.1 would be helpful in this regard, as well.

7.2 ESD Protection and EMI Reduction

Electro-magnetic interference (EMI) can disrupt the circuit's proper operation, or can allow it to disrupt another circuit nearby. This occurs when electrical and magnetic fields are generated and propagated (either conducted through wiring, or radiated through the air). These fields are produced not just by the unwanted electrical noise in a system, but also by the signals themselves. Left unchecked, these fields will produce unwanted noise voltages. Electro-static discharge (ESD) is a phenomenon that can damage electronic circuitry. It is characterized by high-voltage discharges of static electricity that can literally blow apart sections of an integrated circuit.

Both EMI and ESD are undesirable and should be limited as much as possible. There is much that a circuit designer can do to limit a product's susceptibility to both EMI and ESD, and its tendency to generate EMI. Some of the more important strategies are discussed in this section, including proper shielding, grounding, and filtering.

7.2.1 Shielding

The primary protection against both ESD and EMI in any system is provided by the shielding which surrounds it. If designed properly, this *EMI enclosure* traps EMI that a system generates inside it, and it also prevents EMI and ESD generated elsewhere from getting inside.

The shielding is provided by conductive material that makes up or is contained within the mechanical enclosure. The enclosure may be made of aluminum or steel, for example, or it may be a plastic material that has a conductive layer deposited on or embedded in it. Shielding can be effective against both electric and magnetic fields, but magnetic fields are usually more difficult. In either case, some portion of the fields is reflected, and some portion is absorbed. Absorption loss is due to eddy currents that are generated in the shield, and then dissipated (converted into heat) by the shield's internal resistance. Reflection loss is due to impedance variations that the field encounters when striking and passing through the shield (this is the same effect which causes reflections in transmission lines).

The effectiveness of a shield depends on many factors, which include the material used, its thickness, and the frequency of the fields that the shield is attempting to block. Absorption is the primary shielding mechanism for low-frequency magnetic fields. Non-ferrous materials will not provide effective shielding for magnetic fields with frequencies less than about 100 KHz, and even ferrous materials are not very effective against magnetic fields below 10 KHz. For both magnetic and electric fields, regardless of frequency, the material's thickness plays an important part in its absorption effectiveness. Reflection is the primary shielding mechanism for higher-frequency magnetic fields, and for all electric fields. Non-ferrous materials generally work best here, and the thickness is relatively unimportant.

No shield will be effective if it allows EMI to "leak" through its holes and seams. Every EMI enclosure has them, where the case opens, where cables enter or exit, and so on. Interestingly, the severity of a leak does not depend on the area of the hole or seam, but on its maximum dimension (length, width, height, or diameter). A long, narrow, barely noticeable gap where the case closes may actually allow more EMI to pass through it than the seemingly larger hole cut for a cooling fan. Also, a large number of small holes in the shield is much better than one larger hole, even if the total overall area is the same.

The significance of a gap's length, a rectangular cut-out's diagonal measure, or a hole's diameter depends on the wavelength of the frequencies being shielded. Typically, dimensions that are less than about one-fourth the frequencies' shortest wavelength aren't significant. Consider a 50 MHz digital system, which will contain harmonics of at least 200 MHz and above. Shield gaps in such a system must be kept less than about 16 cm (6.3") if the shield is to be effective.

To maximize a shield's effectiveness, then, cut-outs and holes should be kept as small as possible. The shielding material should be overlapped at seams, and electrical gasketing material (such as spring fingers or wire mesh) should be used wherever possible.

PC card designers often do not think about shielding, believing that it is mostly an enclosure design problem. There is much the card designer can do, though. For example, signals with high edge-rates can be routed on layers sandwiched between the board's power and ground planes. This is often called *stripline* routing (*microstrip* is different, because it implies a copper plane on one side, but not the other). Stripline signals are surrounded and shielded by copper, limiting their ability to emit EMI (or be affected by it).

7.2.2 Grounding

In order to complete the EMI enclosure in an SBus system and to provide the utmost ESD protection, backplates need to be grounded. This is easy, right? Simply connect the backplate to logic ground on your board, or rely on an electrical and mechanical connection to the chassis (the backpanel or frontpanel, which must also be grounded). Right?

Wrong.

Unfortunately, proper system grounding procedures are not well understood, and so are often not given the consideration they deserve. A properly grounded system is safer, more reliable, radiates and conducts less electromagnetic interference, and is less prone to damage from electrostatic discharge.

The best way to achieve proper system grounding is to design in and maintain the integrity of the system's ground, part of which is usually the EMI enclosure. Entire books have been devoted to this subject. A complete discussion isn't possible here, of course, but a few key points and pitfalls will be mentioned.

Above all, the system must be designed so that the EMI enclosure, chassis, or other ground system is never allowed to carry current under normal circumstances. The reason for this is simple: any unbalanced current in the EMI enclosure, especially alternating current (AC), is not shielded and can radiate at will. Also, because the impedance of any ground system is finite, currents in the ground system can introduce voltage differences that can detract from signal noise margins.

Generally, the methods for avoiding ground system currents boil down to a few key restrictions. Among these are:

- **Avoid ground loops whenever possible.** A ground loop can be very subtle and difficult to perceive.
- **Do not allow signal return currents to flow through the ground system.** Generally, this is caused by improper cabling, especially improper cable shielding.

Each of these restrictions will be discussed in turn in the following sections.

Ground Loops

A typical situation is shown in Figure 7.13. Two systems (A and B) are connected together via a coaxial cable (C). The electronic core inside each system is tied to the chassis at exactly one point (D and E, respectively). Each chassis is tied to earth ground, probably via its AC cord (at points F and G, respectively).

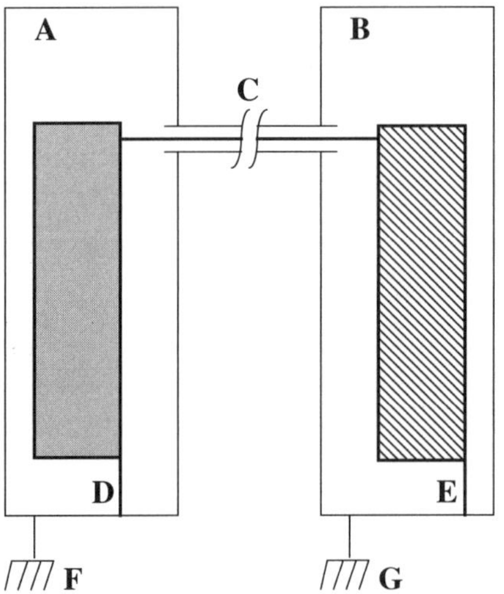

FIGURE 7.13. *Typical Interconnection Situation.*

In this configuration, a circular, electrically conductive path exists, as shown by the dashed line in Figure 7.14. This path is called a ground loop, and is undesirable for several reasons. For example, varying electromagnetic fields in the vicinity will produce currents in the loop, much as the primary of a transformer electromagnetically induces currents in the secondary. Currents can also be induced if the earth grounds are at different potentials, which is common if the grounding is done via an AC outlet and the outlets used are not on the same power trunk or are physically separated from each other. Currents induced via either of these mechanisms can be quite large if the ground system impedance is low (which it should be). These currents obviously can flow through an EMI enclosure.

Note that if the electronic cores in either of these systems were tied to the chassis at more than one point, there would be one or more ground loops contained within the system itself, as illustrated in Figure 7.15. Here, the electronic core in system B is connected to the chassis at two points (E and H), allowing a ground loop current to flow around the dashed path shown.

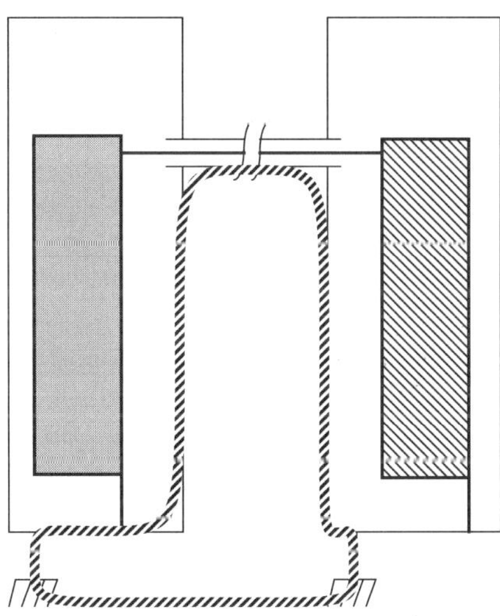

FIGURE 7.14. *Ground-loop Highlighted Through AC Grounds.*

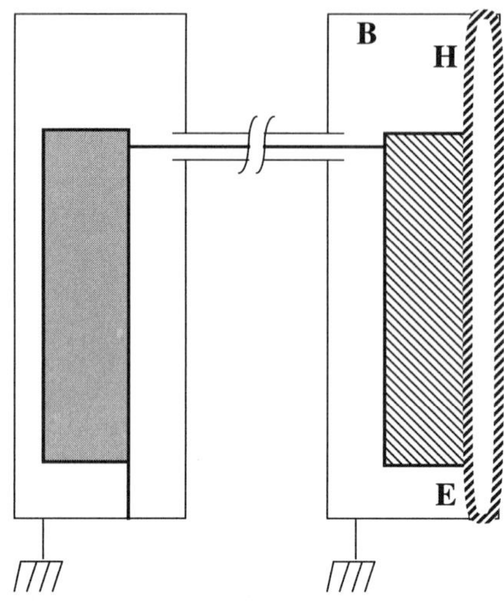

FIGURE 7.15. *Ground-loop Highlighted Through Multiple Chassis Tie-Points.*

Signal Returns

Inadvertent currents can also be induced in the ground system if care isn't taken with signal return paths. Signal return paths for the configuration under consideration are illustrated by Figure 7.16. Notice that the EMI enclosures of both systems are forced to carry the signal return currents. Signal return currents can have very high frequency components, especially for digital and video applications, and significant radiated interference with nearby equipment can occur in such a situation. Even worse, the signal return current could be via the path shown in Figure 7.17. In this case, the path may include some of the actual wiring within a building's walls, forming an antenna of sorts, and likely also leading to interference via conducted emissions.

Recommendations

There are a number of ways to eliminate these problems, especially early in the design phase. Guidelines include:

- First and foremost, try to take a system-wide view of the problem and take alternate configurations into account.

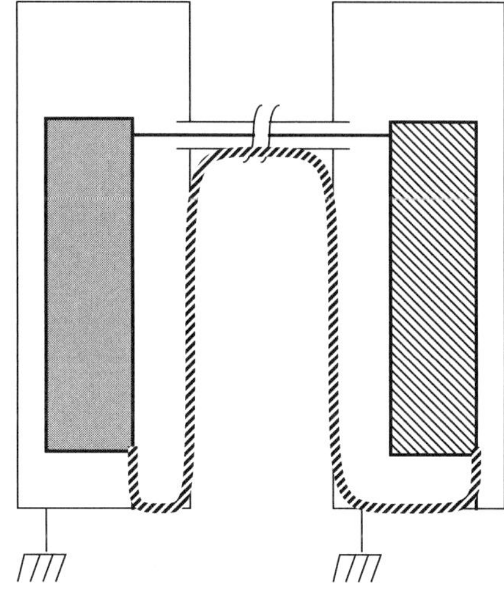

FIGURE 7.16. *Improper Signal Return Path Through EMI Enclosure Shown.*

- Connect logic grounds to chassis grounds at only one point (ideally at the power supply).
- For maximum shielding against magnetic fields, cables should be designed to carry all of the signal return currents.
- For maximum shielding against electric fields, all cables should be grounded in some way, either via ground wire(s) or via the shield. If the cable length is less than about 1/20th of the wavelength of the frequencies in the cable, the ground should be connected at one point only. Otherwise, the cable must be multiply grounded, but extra care is necessary in this case to avoid ground loops.
- Consider driving signals differentially or using transformer or optical isolation.
- Beware of signal grounds (which can easily cause ground loops) in, for example, RS232 cables. Opto-isolated drivers/receivers are becoming increasingly popular for exactly these reasons.

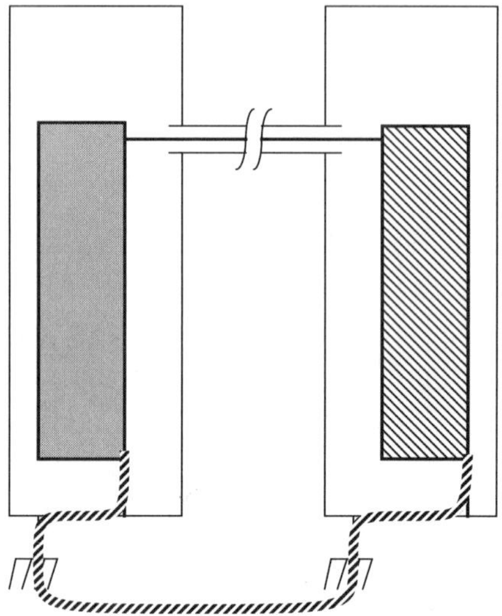

FIGURE 7.17. *Improper Signal Return Path Through AC Grounds Shown.*

- Remember that capacitors (including stray capacitance!) are effectively AC shorts, and are counterproductive when trying to break up ground loops and signal return paths.

Just one of many possible solutions to our current problem is shown in Figure 7.18. Note that the electronic core in system A is isolated from ground. The cable shield is used for the signal return, and is grounded at only one point via the electronics in system B. These electronics are, in turn, grounded to the chassis at only one point.

Backplate Grounding

An SBus card's backplate must not be connected directly to logic ground. When the card is installed it makes physical and electrical contact with the system's EMI enclosure, which is already grounded. If the backplate is mistakenly grounded via the card as well, then undesirable ground loops exist. The shielding effectiveness of the backplate should be maximized if it is grounded *only* through the enclosure.

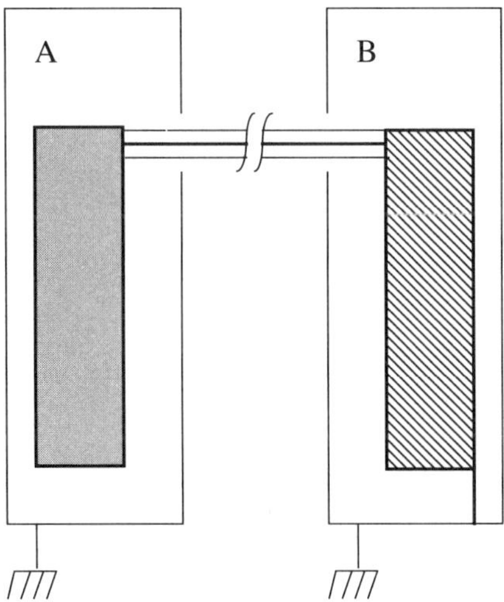

FIGURE 7.18. *One Possible Solution to Both Ground-loop and Signal-Return Problems.*

Grounding and shielding is often something of a "black art," where theory and reality often don't seem closely related. There are several vendors who have reported that grounding their backplates on the card does seem to reduce emissions. Given the previous discussions, how can this be? There is no disputing that this does occur. Does that mean there is an error in the theory, and that backplates really should be grounded?

No, it does not. The theory is correct and is reflected in the measurements and tests taken. The discrepancy lies not in the theory, but in the observation; there are factors involved which the designer does not recognize. These other factors can be as subtle as they are important.

First, an error in the way a system is grounded does not always cause noticeable symptoms, EMI or otherwise. In fact, in many cases there will be no problems whatsoever. It is important to remember, though, that a configuration may change, or a new device may be added, or a different cable connected, or a new host installed. Problems that had been latent may someday arise and cause major havoc. It is better to design such problems out of a system than to have faith you will be able to find and fix them in the lab. It is a very big error to tack on an extra ground connection, make some measurements, and then deduce that the problem is solved.

Why then does grounding the backplate on the card sometimes seem to help? Clearly, for these configurations the benefits of the extra ground connection outweigh the penalties. But why are there any benefits at all, if the backplate is already grounded through the EMI enclosure? The answers here are that the backplate isn't already grounded, or is somehow corrupting that ground. The latter answer is the more likely, for reasons that will become clear.

If the backplate's connection to the EMI enclosure does not have suitably low impedance then it will not prove an effective shield. The backplate should make metal-to-metal contact with the enclosure, preferably on all of its four sides. This metal should be clean and free of corrosion. It should also be unlikely to corrode in the future. Remember, too, that the DC impedance may be very different from the high frequency AC impedance. The latter is most important when working with EMI, and very little inductance in the ground path is required to increase the AC impedance past the point at which grounding is effective.

If the backplate's impedance is reasonable, however, then there is another factor to be considered. Assume that somehow, accidentally, high frequency energy gets coupled into the backplate. This coupling might be either magnetic or capacitive, but in either case the charge that results will flow toward ground. If the backplate is grounded only through the EMI enclosure then this current will flow through the enclosure. Instead of providing a shield, the enclosure will actually become an antenna! Grounding the backplate through the card may help in this case if it provides a lower impedance path to ground than the enclosure. The stray backplate currents will flow through the card, not the enclosure, and any radiated emissions that result will stay inside.

This may seem to be an argument *for* grounding the backplate through the card, but it is not. That is not a good idea for all the reasons previously discussed. It *is* an argument for preventing capacitive or magnetic coupling with the backplate. Coupling due to stray capacitance is the most likely culprit; at high frequencies even very small capacitors (stray or otherwise) are significant. How is the backplate attached to the board? If this is done with brackets, are there any traces or power planes on the board directly underneath the bracket, on any layer? If so, the bracket and the trace or plane form a parallel-plate capacitor with the board's material acting as the dielectric. There are several other possibilities, of course, including split power planes with an insufficient gap between sections.

To test whether or not coupling into the backplate is the root of your EMI problem, remove it! If that is not possible, insulate it so that it does not make contact with the back- or frontpanel. The hole which the backplate is designed to fill is not big enough to allow significant radiation at frequencies even as high as 500 MHz and above (remember that the attenuation a gap or hole provides is a function of its maximum linear dimension, not its area). If an absent or insulated backplate *also* reduces the EMI generated by the product, then it is clear the problem is a caused by coupling into the backplate and the resulting current that flows through the enclosure.

It is possible to properly ground a product and strictly limit its EMI emissions. There are few shortcuts, though. Do not ground the backplate even if it seems to "fix" the problem. You have only succeeded in masking the real problems, while possibly creating others.

7.2.3 Filtering

Filtering noise is an important part of EMI suppression. Filtering helps in two ways. First, it limits noise conduction into the system, where it can interfere with proper operation. Filtering also limits noise conduction out of the system, where it can interfere with the operation of other systems. Both aspects are important. While filtering is primarily focused towards conducted emissions, it can indirectly help reduce radiated emissions (or susceptibility to them). This is because the wires through which noise is conducted can act like antennas.

This subsection concentrates on the problem of filtering digital signals. Noise filtering within power distribution systems is another important aspect of EMI suppression, but that subject is outside of this book's focus.

With the exception of AC "hum," most noise is at high frequencies. Low-pass filters will let the desired signal through, while blocking most noise. One very simple type of low-pass filter is a bypass capacitor, which is connected like a shunt between the signal and ground. The capacitor's impedance decreases with increasing frequency, acting like a virtual short circuit for noise while hardly affecting lower frequency signals. The effectiveness of this capacitor can be increased by including one or more resistors. Three common configurations are shown in Figure 7.19. The simplest of these uses only one resistor and one capacitor, but it filters best in only one direction. The other two configurations shown filter equally well in either direction. Assuming that the source impedance is small compared to the resistor value, all of these filter configurations are *first-order* circuits. This means that above the filter's cut-off frequency, the response rolls off by about 6 dB per octave (two frequencies differ by an octave if one is twice the frequency of the other). Roll-offs sharper than this can be achieved by using multiple stages in series.

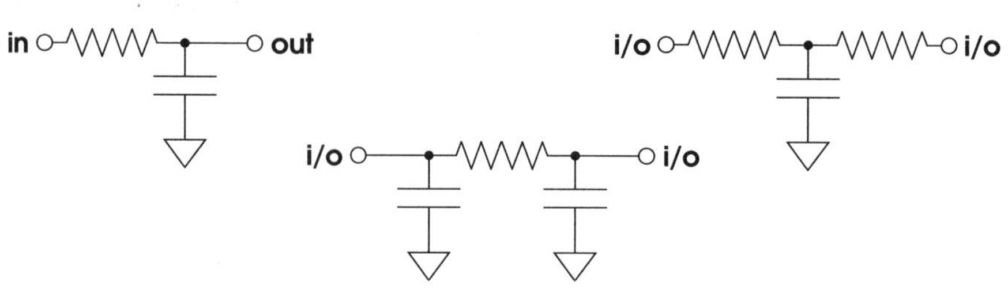

FIGURE 7.19. *Three Possible R-C Low-Pass Filter Configurations.*

Low-pass R-C filters are available in Single Inline Package (SIP), Dual Inline Package (DIP), and surface mount packages in any of several different configurations. These can be useful for filtering many different signals simultaneously. For example, every signal in an RS-232 interface could be filtered using such packages. When filtering a digital signal, it is important to remember that the signal edges are composed of harmonics whose frequencies are integer multiples of the base signal's frequency. To avoid excessive degradation of signal edges, then, pick the filter's cut-off frequency so that it is at least three to four times the frequency of the desired signal. If filtering a 19.2 Kbaud RS-232 interface, for instance, an 80 KHz filter would be a good choice.

Low-pass filters can also be made using inductors. Their impedance increases with frequency; higher frequencies are attenuated when passing through an inductor, while lower frequencies are relatively unaffected. When combined with capacitors (or a trace's stray capacitance), inductors can be very effective filters. Inductors are available in many forms, including surface mountable "chip" inductors. Like R-C filters, L-C filters are also available pre-packaged and ready to use. An example of one such filter in a SIP package is shown in Figure 7.20. This is a fifth-order, low-pass filter, which rolls off by about 30 dB per octave. This gives very sharp attenuation to any noise above the desired cut-off frequency.

Ferrite beads and toroids are another alternative. These are little "donuts" of magnetic material. When a conductor is threaded through or wrapped around one its inductance is enhanced, filtering high frequencies as described above. These may not sound like they are very compatible with high-volume printed circuit board production, but in fact they can be. Ferrite bead filters are available in pre-fabricated axial forms that can be mounted to boards much like resistors are. They are also available in packages that look much like integrated circuits, and can filter several different signals at once (sometimes using the same ferrite structure; this can provide good common-mode filtering, although it may increase low-frequency cross-talk slightly).

Whatever the method used to filter signals, it is important that inadvertent bypasses are not built into the design. For example, stray capacitance can provide a convenient mechanism for coupling noise back into a system, as can cross-talk. Keep filtered and un-filtered signals separate. This can be done with sufficient physical distance, or by providing shielding (using an intervening ground plane or low-impedance trace).

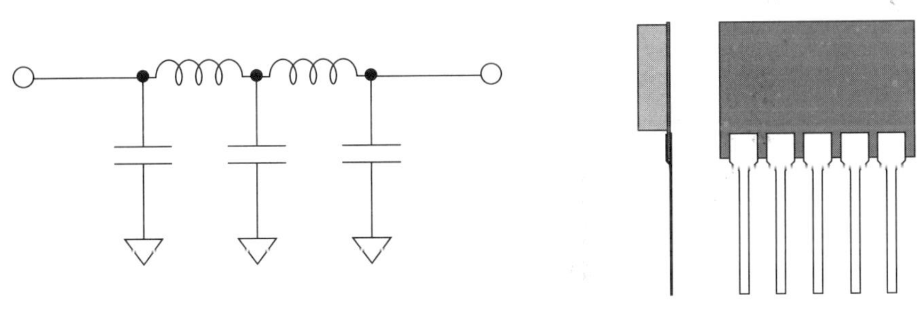

FIGURE 7.20. *Fifth-Order L-C Low-Pass Filter Component.*

7.2.4 Regulatory Agency Compliance

Worldwide, there are many different agencies that regulate safety and EMI standards for products. For product safety, the agencies involved include UL (USA), CSA (Canada), and TUV (Europe). For EMI, the agencies include FCC (USA), DOC (Canada), VDE (Europe), and VCCI (Japan). All product manufacturers are responsible for insuring compliance with all regulations applicable wherever the product is sold. This includes SBus add-on cards, which are usually tested in a "representative" host.

The vendor may be legally responsible if any safety or EMI related problems arise with the product, and so it is wise to seek and secure agency certification. This can be a complex process. If the vendor does not have the required expertise on staff, it may be helpful to hire a consultant.

References

Ott, H. W. *Noise Reduction Techniques In Electronic Systems*. John Wiley and Sons, 1976.

Barnes, J. R. *Electronic System Design Interference and Noise Control Techniques*. Prentice-Hall, 1987.

SDS Data Systems. *Grounding and Noise Reduction Practices*.

Getting Your Hands Dirty

8

For those considering the design of a product which uses the SBus, eventually it comes time to actually design, build, and test it. This chapter's aim is to help make that effort as successful as possible. First, inter-operability issues that must be considered early in the design are presented. Then a checklist and some guidelines are provided; the former for use when reviewing a design for completeness, and the latter for use when preparing to apply power to the prototype for the first time. The final section, though, may be the most useful. It provides troubleshooting clues which may prove helpful in isolating and correcting problems that appear when trying out the new design.

8.1 Inter-Operability

A compliant design means one that rigorously adheres to all applicable specifications. A compatible design is one that will work, whether or not it is compliant. Two compliant designs may not be compatible unless the specification is complete and unambiguous. Conversely, two compatible designs are not necessarily compliant; they may both deviate from the specification as long as those deviations do not prove fatal to the system as a whole.

Ultimately, the customer cares most about compatibility. He or she is less concerned with the technical details of *how* a device works than with *whether* it will work once it is installed. In simple circumstances it may be possible to produce a list which details which products will work with each other, and this will be sufficient. Any new product introduced can be tested with all other devices on the list, which is then revised accordingly. This might be the approach taken by a company with a (usually proprietary) product line that has relatively few elements and configurations.

Such an approach falls apart rapidly when the product line is diverse, with many elements and interdependencies. The situation is even worse when a variety of independent vendors are participating, and there is no single cohesive strategy involved.

Further complicating matters, such compatibility testing is usually done empirically; by observation and experiment. There is often no guarantee that the results of such testing represent worst-case scenarios. In the case of something like the SBus this might mean that a board might work in a machine, but only if this vendor's parts are used and that vendor's card is not plugged alongside, etc. Simple process variation in an otherwise identical design might mean the difference between a compatible board, and one that is not.

Compliance is a mechanism that can simplify the problem of insuring compatibility. If the interface is well defined then it should be possible to build a compliant board that has a high probability of being compatible with all other compliant designs. No specification is ever perfect and hence compliance is no guarantee, but it can be much simpler and provide a higher degree of confidence in the long run than compatibility-only testing can.

While specification compliance is a very important goal, it is not enough in and of itself. That is because not every product in the installed base is compliant; they contain errors or may be compliant with a previous revision of the specification. Therefore, a good design will strive not just to be compliant, but also to maximize its ability to be compatible with non-compliant or non-current products.

The purpose of this section is to help promote compatibility and remove as many configuration-related concerns as possible. Maximizing compliance with the specification is one important aspect of this, and information and guidelines are provided to help insure that. Also important are some key issues which relate to products which are non-compliant or have design flaws, but are prevalent in the installed base. Many of these were discussed in Chapter 5, but others (those relating to design features and decisions for new products) are included here.

8.1.1 SBus "Profiles"

For other buses, "profiles" divide the specification into varied classes or categories. Each of these categories might have a unique form-factor, or subset of features, or both. Most specifications contain optional features and forms, and profiles are a mechanism used to narrow the choices that a designer has to make. Rather than picking and choosing among a wide variety of features, he or she picks an appropriate profile and designs to its requirements. The advantage of this approach is that all products within a given profile are more likely to be compatible with one another. It also allows a more general "mother" specification which can better suit a wide variety of applications.

Although the original intent of the SBus "profiles" was similar, the term has come to have a different meaning. SBus, as defined in revision B.0, already has a mechanism for managing variations in configuration: the "Open Boot PROM" firmware architecture described in Chapter 4. As a result, SBus profiles became a definition of the minimum feature subset which is guaranteed. This is needed for situations where Open Boot does not exist or is not yet functioning.

Any card which cannot or does not interface to Open Boot must be built along the SBus profile guidelines. Any other SBus features may be used only if the card successfully "negotiates" with Open Boot and determines that feature will work or can be used in this host. Host designers must provide support for at least the minimum profile requirements, and are encouraged to include Open Boot in their firmware architectures.

SBus profile guidelines require that at least 25 physical address bits (PA(24:0)) are provided by the host, and transfer sizes of 1, 2, 4, and 16 bytes must be possible. Data parity should not be checked, and extended mode (64-bit) transfers should not be used, unless specifically enabled.

Documentation on the Open Boot PROM interface is concentrated in the documents *Writing FCode Programs for SBus Cards*, and *Open Boot PROM Toolkit User's Guide*, published by Sun Microsystems (see this book's bibliography). In essence, Open Boot contains pre-defined "attributes" which specify features and capabilities contained within this particular host. attributes are defined for things like SBus clock rate, the number of physical address bits provided, supported burst sizes, and so on. The FCode within a card can query these attributes to determine what features are supported and what behaviors are required.

The list of defined attributes will vary from one machine to the next, as new definitions are added and certain others aren't applicable in all cases. Therefore, a card's FCode must be written so that it will make a default decision if an attribute can't be found. That is where SBus profiles come in handy. The name may not be entirely accurate (and may change in future revisions of the specification), but the purpose is critical.

8.1.2 Hardware Issues

This subsection contains various discussions about hardware issues which affect inter-operability.

Avoid Restricting Acknowledgments

Revision B.0 of the SBus Specification allows masters to restrict the types of acknowledgments which it accepts, and treat all others like error acknowledgments. In other words, the master may choose not to be able to perform transfers with slaves whose port-widths don't match the particular subset that the master supports.

This permission was granted because it allows simplifying trade-offs to be made in the master's design. In particular, byte-steering and bus sizing can be simplified or eliminated altogether.

The obvious consequence of these trade-offs, though, is that some SBus masters cannot transfer data to or from some slaves. The most common assumption used to justify these trade-offs is that, "My device only transfers data to and from system memory, and I know how that behaves. Why should I include extra complexity to account for circumstances that I know won't ever arise?" The problem with this argument is that it places too much faith in the notion that the installed base's behavior is well-defined now, and won't change appreciably in the future. This isn't necessarily true.

Ideally, any combination of SBus masters and slaves could transfer data. One way for the master to help insure this is to avoid restricting the acknowledgments which it accepts; support slaves which are 8-, 16-, or 32-bits wide. While this is only a strong recommendation right now, future revisions of the specification will probably make it a requirement.

Avoid Restricting SIZ Codes

If designing an SBus slave interface, avoid restricting which **SIZ** codes are legal. If the slave port is 8 bits wide, respond with a byte acknowledgment to either byte, half-word, or word accesses. If the slave port is 16 bits wide, respond with a half-word acknowledgment to either half-word or word accesses. If the transfer size encoded is greater than the acknowledgment received then the master may resort to bus sizing, at its discretion. No additional byte steering or other logic is required on the slave. That is the responsibility of the master during bus sizing.

The key point here is that the slave does not benefit from unduly restricting which size codes are acceptable. Offering the master a bus sizing option costs the slave little, if anything, and it can greatly increase the number of different interconnections possible. This, in turn, maximizes compatibility.

Keep in mind that some SBus hosts probe the SBus slots first with 32-bit accesses. If the ID PROM is connected to a byte-wide port which responds to 32-bit accesses with error acknowledgments, then the board will be ignored for the remainder of the probe sequence, or the machine will fail to boot and seem to hang. For more information on this see Section 4.2.1 on page 116.

Determining Optimum Interrupt Levels

An SBus card may use one or more of the SBus interrupt lines to signal the host processor that some service is needed. These lines are prioritized, from one (the lowest) to seven (the highest). Picking the right priority is important. If too low a priority is used then the device may not be serviced often or quickly enough. If too high a level is chosen then this device will probably work great, but some other device may not get the attention it needs. In either case the end result will usually be a system crash and an unhappy customer.

Picking the right interrupt level can be difficult, though. System configurations have almost limitless variations. There is no way for a developer to know exactly what type of host their card will be plugged into, or what other devices it will have to share the system's resources with.

One solution is to force the end-users to make the decision. After all, they know much more about the configuration of their machines, what software they use, and what performance they need. They may not know enough about your design (or even about their machines) to make an intelligent decision, though. Also, they may resent the complexity required to install the card, and to reconfigure it occasionally when a new device is added or a different application is used. Ideally the end-user should not have to set a switch or insert a jumper, as this is counter to SBus' auto-configuration goals. IBM-compatible personal computers require the user to do this, and it is a major source of difficulty and a common cause of compatibility problems.

A better solution might be to design the SBus card in such a way that it can electronically select the appropriate interrupt level. This adds complexity and cost to the final product, though, and it is difficult to make the corresponding adjustment to the **intr** FCode parameter contained in the board's ID Prom.

A much better solution for all concerned is to find and fix an interrupt level (or levels) that can be expected to work across all configurations. This can be done with some thought, and with careful design of both the card's hardware and its driver.

Fortunately, the exact priority level chosen is usually not that critical. Most often there is a range of priorities that will give adequate results. Also, SBus interrupts are easily shared because they are driven by open-drain (or open-collector) drivers and are level-sensitive.

When designing the card, it is important to remember that interrupt service latency cannot be guaranteed even in the best of circumstances. Interrupts are intended to be used to awaken the device's driver after some condition has been met or some action has been completed. One example of this would be a card that issues an interrupt when a block of data has been received, or when a co-processor has completed an extended calculation and is ready with a result. It is generally inappropriate to use interrupts to transfer data on a byte-by-byte or a word-by-word basis, unless the data rate is very slow. In a design such as this it is better to buffer the data and signal the processor only when the buffer is half-full, for example. Alternatively, DVMA operations can be used to transfer the data directly to memory, interrupting the processor only when the transfer is done. Both of these methods reduce the design's sensitivity to interrupt priority level, and they are also much more efficient use of the SBus's data and interrupt service bandwidths.

If your card can generate interrupts under a variety of different conditions, then it may also be beneficial to use multiple interrupt lines. High priority interrupt requests can then be signaled and serviced separately from lower priority requests. Otherwise, all interrupts would be serviced at the same level needed by the highest priority event. This means that some events are serviced at a higher priority than necessary, and this is wasteful.

When narrowing down the proper priority for your device it is important to consider it relative to other devices that may be present in the host system. Some general questions are helpful here. First, what happens if the interrupt does not happen immediately? Does data get lost, or some other major failure occur? Or does performance simply degrade? If so, how much? While a performance loss is certainly not desirable, it is much better than a failure. It is generally better to give such devices a relatively low priority, because they survive if temporarily pushed aside by a device whose accesses are more critical. Co-processors tend to fall into this category because it is usually not a problem if a result isn't picked up as soon as it is ready.

Another question to ask is the expected frequency of the interrupts. Are they frequent? Few and far between? Do they tend to occur in bursts, or are they either random or regular? Infrequent interrupts can often be serviced at a level higher than absolutely necessary without much of a penalty. Interrupts that are frequent or that occur in bursts, though, can be a significant problem unless care is taken to service them at as low a level as possible.

Consider a situation where a small group of devices must be given relative priorities. The first device is an unbuffered serial port that interrupts whenever a character is received. The second device is a floppy disk controller which is also unbuffered and which also interrupts whenever a decoded byte is available. The last device is a digital signal processor that interrupts the processor whenever it has completed one calculation and is ready for another, or whenever new data is needed before it can proceed. This data is transferred efficiently in blocks, but a lot of data is needed and hence many blocks must be transferred.

The data rate on even a very fast serial port is low compared to the overall bandwidths in most SBus machines. Floppy disk transfer rates are usually quite a bit higher, although still relatively modest. Digital signal processors, though, can be very fast, consuming large amounts of data often as fast as it can be delivered. Customers that buy DSP based products are often most interested in the device's performance, and want as much "bang-for-the-buck" as possible.

At first glance it may seem that the serial port in this example can be given the lowest priority. Next comes the floppy disk, and then the DSP card can be given the highest priority. This priority ranking is wrong, though, and might result in an unreliable system.

If the serial port is not serviced often enough, then the data it receives will be overrun and lost forever. Besides, the relatively low data (hence interrupt) rate would not pose much of a burden on the system even if serviced at a higher then necessary rate. This device's priority should be relatively high.

Data overruns can occur in the floppy disk, too, because it is also unbuffered. In this case the data isn't lost forever, though, because it can be read again the next time it rotates past the disk's read heads. This is a big penalty to pay, because the time consumed in just one disk rotation is large. Still, the error is not fatal. This device's priority should be moderate.

The digital signal-processor won't lose data if it is not serviced quickly enough. It may temporarily stop processing because it is starved for data, or because it is unable to pass off a result, but this is not fatal. Neither is the penalty associated with this temporary stall very high in most cases. The DSP can restart as soon as the host is able to service the interrupt; it is not necessary to wait for a mechanical disk rotation or for any other long term event. The DSP will interrupt frequently, too, which might easily interfere with other devices if serviced at too high a priority. Obviously the customer will want to squeeze everything possible out of his DSP investment, but probably not to a degree that it adversely effects the reliability or efficiency of other parts of the host. This device should be given a relatively low priority.

In this example, then, the serial port should be given the highest priority, followed by the floppy diskette interface. The DSP card is given lowest priority. Note that it might be possible to reduce any performance impact on the DSP by giving different priority levels to those interrupts caused by data requests (frequent) and to those generated when results are available (relatively infrequent). The former should remain at a low priority, but the priority of the latter can be increased.

In this example device priorities have been set relative to other devices whose type and characteristics are known. Usually the designer does not know what other devices his or her board will have to coexist and cooperate with. The following two tables can help to pin down the appropriate SBus interrupt levels, though.

SBus-based machines normally contain on-board devices, such as serial ports, video interfaces, ethernet ports, and SCSI ports. The designers of these machines have had to hard-wire the priority of these devices, and factor in the SBus interrupts, and other interrupt sources (both hardware and software) as well. In effect, much of the ranking has been done for you. If your device is similar to those already listed then the same interrupt level is a good first estimate. If the device doesn't map exactly to an SBus interrupt level then pick the closest one that does. If your device does not closely correspond to one of those devices listed, then attempt to find the proper relative position using those techniques described above. In all cases it is better to pick the lowest level that will provide an adequate interrupt service rate. Picking a higher level than necessary isn't just wasteful and potentially harmful to your "neighbor" devices. It may also cause compatibility problems. See the related discussion on page 186.

The SPARCstation 1, 1+, 2, and 2+ interrupt priorities are mapped as shown in Figure 8.1. This table will be typical of single processor desktop systems. It may be surprising that the ethernet and SCSI interfaces seem to have overly low priorities given their data rates. These on-board devices transfer their data with DVMA

SBus	CPU Level	Usage
	15	Asynchronous Memory Error
	14	Counter #1
	13	Audio
	12	Keyboard, Mouse, Serial Ports
	11	Floppy Diskette
	10	Counter #0
7	9	
6	8	
5	7	Video
	6	Software Interrupt
4	5	Ethernet
	4	Software Interrupt
3	3	SCSI
2	2	
1	1	Software Interrupt

FIGURE 8.1. *SPARCstation 1/1+, 2/2+ Family Interrupt Level Mapping.*

operations that do not require interrupts. The interrupts are only necessary to set up or finish a block transfer.

In multi-processor systems, or systems that have VMEbus, Futurebus+, and/or MBus interfaces in addition to the SBus, the interrupt priority mapping will be more like that shown in Figure 8.2. This table represents the priority mappings found in SPARC-server 600 MP class machines. As mentioned, the relative ordering is consistent, although the overall priority levels have been increased to reflect the greater memory and interrupt service latency expected in these machines.

Special care should be exercised when choosing SBus interrupt levels 6 and 7. Please see the discussion on page 186 and the related section that starts on page 184 for more information.

Notice that despite the variation in these tables, the relative device ordering remains fairly consistent. These tables are not the only mappings possible. The SBus Specification does not restrict this in any way. These tables are expected to be fairly typical, however.

Notice also the software interrupts in both of the preceding tables. This is a very useful mechanism, and can be used by a device driver to choose whichever priority is best for it on the fly.

The device's hardware interrupt level cannot be easily changed once it is set, but that level can be used only to schedule a software interrupt. This software can be of any priority level less-than or equal to the original hardware level. Please see the discussion titled 'Interrupt Mappings Might Differ,' starting on page 186 for more information.

SBus	CPU Level	Usage
	15	Software Int., Async Mem Error
	14	Software Int., Proc. Counter
7	13	Software Int., VMEbus 7, Audio
	12	Software Int., Serial Ports
6	11	Software Int., VMEbus 6, Floppy
	10	Software Int., System Counter
5	9	Software Int., VMEbus 5
	8	Software Int., Video
4	7	Software Int., VMEbus 4
	6	Software Int., Ethernet
3	5	Software Int., VMEbus 3
	4	Software Int., SCSI
2	3	Software Int., VMEbus 2
1	2	Software Int., VMEbus 1
	1	Software Int.

FIGURE 8.2. *SPARCserver 600 MP Family Interrupt Level Mapping.*

8.1.3 Open Boot PROM Rev 1 vs. Rev 2

The biggest issue regarding firmware inter-operability focuses on the differences between Open Boot PROM Revisions 1.X and 2.X. The former was shipped with SPARCstation 1/1+ machines, and other early machines which share that architecture. The latter is the more current (and the more capable) of the two. This subsection focuses on some of the key differences between the two that can most affect inter-operability.

"Name" vs. "Driver" Attributes

The **driver** FCode has slightly different behavior in 2.x revisions than it had in 1.x revisions. This difference will affect only those cards which use this FCode, and with an associated text string that does *not* start with exactly four characters followed by a comma.

When processing the **driver** FCode, 1.X PROMS strip exactly five characters from the front of the associated text string and build the device's name from the rest. On the other hand, 2.X PROMS strip characters up to and including the first comma. As a result, if the text string's fifth character isn't a comma, the device's name won't be correct in 1.x environments.

The **driver** FCode should not be used due to these problems, and because future support for it in any form is not guaranteed. It has been replaced by the **name** attribute, which has no known compatibility problems, and is the preferred way of identifying the device.

"Finish-Device" FCode Compatibility Problems

The **new-device** and **finish-device** FCodes are not needed for SBus cards which contain only a single device, and in fact should not be used in this case. Versions 1.X of the Open Boot Prom contain place-holders for these functions which are little more than no-ops; if used nothing will happen, useful or otherwise. No errors will occur even if these FCodes are used improperly.

Open Boot PROM versions 2.X and above do fully implement these functions, though. Any card which improperly makes use of the FCodes will fail in this environment, even though it *appeared* to operate properly before.

The fix is simple; remove these FCodes if they are not needed. If they are, make sure that they are used properly.

8.1.4 Compatibility and Compliance Testing

Until recently an SBus consumer could only rely on the vendor's integrity when seeking assurance than an SBus product was indeed SBus compatible and compliant. Now though, the size and breadth of the SBus market has fostered an interest in testing services which can independently determine just how well an SBus product has been implemented. As of this writing at least two organizations have sought to fill this need. The approaches taken are different, but complementary; one emphasizes compatibility, and the other compliance.

SPARC International

SPARC International is a confederation of many different companies that have aligned themselves with the SPARC microprocessor architecture. The member companies are independent from one another, and in fact are in many cases direct competitors. The purpose of this alliance is to promote the use of SPARC in the marketplace, and in that way to maximize the market's size.

SPARC International has focused a major part of its resources toward standardization, which helps the market grow by promoting inter-operability and end-user confidence and satisfaction. SPARC International tests SPARC-based hardware and software products for compliance with the SPARC specification. If the product passes, then it is allowed to carry a sticker or label which indicates that.

Now, SPARC International has branched out into SBus testing. This may seem odd, at first, because the SBus is not dependent on the SPARC architecture. SBus interfaces can be attached to machines which use any other type of processor. Further, SPARC-processor-based machines can use any other type of interface. Some have already been built around the VME and ISA buses, for example.

Despite the name, though, SPARC International's move toward SBus testing is a logical one. SPARC and SBus are not dependent on one another, but the combination is a good one and increasingly, SBus is becoming the de-facto standard on SPARC-based machines. As a result, SBus standardization and testing can help many of SPARC International's members in the same way that SPARC standardization and testing does.

As part of their testing, SPARC International has collected samples of many different SPARC- (and now SBus-) based products in their laboratories. New SBus products can be tested against these. After the testing, the product vendor will be supplied with a report that details which products have been shown to be compatible.

As might be expected, this is a compatibility test, and it cannot guarantee compliance with the SBus Specification. It does provide very useful information that cuts directly to the point that compliance testing strives for: an assurance that any new product will work with those already in the installed base.

VME Laboratories

VME Laboratories does provide true compliance testing. As the name implies, this company's original specialty was VME compliance testing.

This company employs a two-pronged approach to true compliance testing. The first is an exhaustive design review process. Engineers study the design from the bottom up using a rigorously defined methodology. The advantage of this is that it can provide true worst-case results. Responses to boundary conditions can also be predicted, as can failure modes.

Afterwards, boards are instrumented and tested to confirm results predicted during the design review. Electrical measurements are done, too. Unlike the design review, though, this part of the testing relies on a limited sample, and can't guarantee worst-case results.

The testing that VME Laboratories performs, and the certification that it can provide, focus only on the bus interface. The remainder of the product remains untested and is largely unused. Once given, though, the certification is a very strong indication that the product does conform to the interface specification.

8.2 Design Review Checklist

This section contains an extensive checklist meant to help guarantee that an SBus design is thorough, robust, compatible, and compliant. This checklist should be used before, during, and after the detailed design of an SBus product. Also, it is usually helpful to ask for help in the review process. The engineer responsible for a design may be too close to it to be able to step back and provide an objective review. Often someone who has not been directly involved can provide fresh perspectives and valuable feedback, all of which can improve the design's overall quality.

Some of these checklist items pertain to SBus masters, some to controllers, and still others only to slaves. Most are aimed primarily at SBus add-in cards, but some are intended for SBus host products. When reviewing the checklist, the designer will need to first determine which questions are applicable to his or her design. Fortunately, many of the questions are worded to help make this decision easy. This checklist is also worded so that the designer of a compliant product should be able to answer, "Yes," to all applicable questions.

This list is not meant to be all-inclusive. In any design review there will be questions and concerns that go far beyond what is contained here. These questions center only on the SBus interface, and are focused mostly on areas that are problematic.

Specification Related Issues

- Is the level of the SBus Specification used revision B.0 or later?
- Are there no known cases where the specification has been violated?

Technology and Electrical issues

- Is the design strictly CMOS based? If not, are all components which interface directly to the SBus CMOS?

- Do the electrical logic levels conform in all cases?

- Are leakage currents, capacitance, and stub lengths within limits?

- If building an SBus host whose bus clock rate is greater than 20 MHz, total capacitance limits are reduced to 100 pF. If applicable, is this more stringent requirement satisfied?

- Are interrupts driven only by open-collector (open-drain) or equivalent outputs?

Timing Issues

- Are synchronous signals sampled only on the rising edge of the clock?

- Does the design avoid using the clock's falling edge?

- Has timing analysis been done with rise and fall time delays included?

- Is the card capable of working properly across the entire range of SBus clock frequencies?

- If building a host, is the clock skew within limits? Over the entire range of clock load capacitance expected?

- If building a card, is the Clock input's capacitance less than 20 pF (required by the specification)? Is it greater than 12 pF (this isn't required, but it's a good idea; see Section 5.2.5)?

Power-related Issues

- Does the card never exceed the allocated power dissipations under any circumstances? Is the peak inrush current within the specified limits? Is the average within the specified limits?

- Is the card designed so that it is not possible for any external device (such as an ethernet MUX box or a SCSI terminator) to cause more power to be drawn (either by the card, or by the device, or both) than that allocated?

- Does the card limit solder-side power dissipation to 2 watts or less?

Protocol Issues

- Will the card function properly if the **Reset*** signal is in a high, low, or indeterminate state when power is first applied? (It will become asserted for the required minimum number of cycles, but its initial state is not guaranteed.)

- Are **AS*** and **SEL*** always used synchronously? Is one always qualified with the other?

- Does the slave issue an error acknowledgment for all unsupported sizes?

- Does the slave avoid issuing an error acknowledgment on write transfers of any kind? If not, have you a work-around for the write time-out bug (See coverage starting on page 155), and the OBP write bug (See coverage starting on page 183)?

- Does the device avoid using SBus interrupt levels 6 or 7? If not, do you understand the possible problems this may cause with some streams drivers on SPARCserver MP and similar machines?

- If designing an SBus master, does it base all its (non extended mode) sequencing only on **BG***, and not **AS***?

- Does the master or slave always transfer data in the proper byte lane(s)? This may require the master to drive multiple copies of write data onto a variety of byte lanes, and it may require the master to multiplex read data from a variety of byte lanes.

- Does the master always retry the identical transfer after receiving a rerun acknowledgment, without any other intervening transfers? See the SBus Specification, Revision B.0, page 68, first and second paragraphs.

- The SBus' physical addresses are byte addresses. Does the half-word- or word-wide slave correctly 'ignore' the least significant one or two address bits, respectively? For example, a card which contains a half-word wide memory array 2^n half-words deep should address the array with address bits $(n:1)$, not $((n-1):0)$.

- Does the slave avoid generating error acknowledgments on (perceived) address alignment errors?

- Are atomic operations avoided? If not, have you investigated all other alternatives? Have you thoroughly researched the possible compatibility problems that may result?

- Are the **Ack(2:0)*** signals actively de-asserted for one cycle after they have been asserted?
- Is it impossible for the slave to issue a byte or half-word acknowledgment during any burst transfer? Is it also impossible for the slave to issue a word acknowledgment during any extended-mode burst transfer?
- If the slave is capable of participating in burst transfers, does it correctly perform address wrapping for all acceptable burst transfer lengths?
- If you are building a host, does the controller recognize and support all sizes of burst operations?
- Does the controller allow a master to perform transfers with its own slave?
- Does the controller avoid restricting access to a slave that has given a rerun acknowledgment? Will it allow any master access to the slave, even if it is not the one that received the rerun acknowledgment?
- If you are designing an SBus Controller, does it guarantee that all **BG*** signals are simultaneously de-asserted between transfers? This is necessary to prevent the current transfer's data and the next transfer's virtual address from overlapping.
- If building a host or bus bridge, does it maintain the order of SBus transfers (and interrupts)? See the related discussion which starts on page 164.
- Can the slave port still be accessed if the master port is waiting to perform a transfer? (In such circumstances it is acceptable to generate rerun acknowledgments for a small number of access attempts. It *must* not do so indefinitely, however, or deadlock problems may result.)

Extended Mode Issues

- Does the master drive the Extended Transfer Information Word onto the D(31:0) lines during the cycle following that in which it drove the virtual address onto those lines?
- Does the master actively drive the **Read** signal low during the same cycle in which the Extended Transfer Information Word is driven?

Mechanical Issues

- Does the card use the 2-piece backplate design?
- Does the card's printed circuit board contain two holes which straddle the SBus connector? (these are used for the SBus retainer, stand-off, or other board retention mechanism). Are these holes placed asymmetrically, with one of them closer to the board's edge?
- Is the backplate isolated from logic ground?
- Does the host's design guarantee the required gap under SBus add-in cards?
- Are the trace stub lengths within limits?
- If designing an SBus card, is it a single-width or (at most) a double width card?

Firmware issues

- Does the ID Prom respond to both 8-bit and 32-bit accesses?
- If the card contains multiple register sets, are they all defined within a single **reg** attribute declaration? (See *Writing FCode Programs for SBus Cards*, page 21).
- If the card uses multiple interrupt levels, are they all defined within a single **intr** attribute declaration? (See *Writing FCode Programs for SBus Cards*, page 21).
- Will the FCode work properly with Open Boot PROM revisions 1.X and 2.X?

Software Issues

- Does the device's driver allocate and then free DVMA mappings on a transfer-by-transfer basis?

Profile Issues

- If a DVMA master is capable of performing burst transfers, can it be programmed so that it will only perform 16-byte bursts?
- If a master is capable of extended mode (64-bit) transfers, can it be programmed to perform only standard mode (32-bit) transfers?

Other Issues

- Does the card's design avoid DIP switches or jumpers that must be set prior to installation? If not, have you considered all possible alternatives which are more automated?

- Has the host design (hardware or software) considered the effects of very long rise times on the **IRQ(7:1)*** signals?

- Do the lowest 64 Kbytes of the card's address space contain read-only devices? (See section 3.4.5).

8.3 Pre-Test Guidelines

Once a design has been completed and either prototyped or manufactured, the "moment of truth" is at hand. The time has come to plug it in, turn it on, and "look for smoke". This can be a nerve-wracking experience for many engineers, especially when you're hovering over a workstation or server worth $10,000 or more.

Now suppose that you turn it on, don't see any smoke, but don't see any signs of life, either. The CRT stays blank and the machine just lays there, open and exposed; not even the Toolkit's **ok** prompt to help you figure out what's wrong. Moments like these are often the worst ones in an engineer's life. Right away you know that the rest of this day (and probably week) are going to be spent huddled over an oscilloscope or logic analyzer.

The purpose of this section is to help avoid this kind of scene. A brief checklist is provided that can be used before powering-up a new SBus product for the first time. This might help prevent some of the simple mistakes which are so commonly and universally made on the first prototypes of a design.

8.3.1 Mechanical Inspection

Printed circuit boards should be inspected both before and after the components are attached. A large number of the mistakes that can be made when fabricating or assembling it are visible to the naked eye. These can often be eliminated *before* they have a chance to make smoke, just by carefully examining the board.

Printed Circuit Board Quality

After the board has been fabricated, check the quality of at least a few samples before sending them off to be "stuffed." Look for evidence that the layers are aligned; hold the board up to a bright light, check for power to ground shorts, and so on. Check also for over- or under-etching, and carefully examine the quality and alignment of the solder mask. Also check the silkscreen. Does it contain enough information to help the board assemblers? Are any component ID numbers positioned in a way that makes it unclear where or how the component should be inserted?

Assembly Quality

After getting a board that has just been assembled, carefully check its quality, and use a magnifying lens if one is available. Are the solder joints good? Is there too much or too little solder anywhere? Look especially for solder-bridges on the pins of any fine-pitch or surface-mount components. Then hold the board at an angle, with the solder-side up. Move and tilt it back and forth until the light shines off the pins protruding through the board. Do any appear to be missing, or too short?

If some of the board is hand-wired (if it has bug-fixes or is a prototype, for example), do the wires look solid? Has any of the insulation been scraped or burned away?

Does the board mechanically fit, into either an SBus slot or (if it's a motherboard) its enclosure? If the fit is difficult or tight, don't force it! There may be a design error, and not just a stubborn connector or bent backplate.

Component Orientation

Component related mistakes are very common on early board assemblies. It is beneficial to check the board against a bill of materials, and check the bill of materials, too; it may contain errors.

Are all the components in place? Look especially at the socketed components; are they and the components oriented correctly? If the design contains PALs or PROMs, are they programmed? Correctly? Are the correct CMOS variants used, or has someone mistakenly used a bipolar part because they assumed it was equivalent?

Check the pins on all parts, too. Do any appear to be folded under or broken off? Are all the bypass capacitors, and pull-up and pull-down resistors in place?

8.3.2 Power Tests

After the mechanical inspection, the board's power-to-ground impedance should be checked. This is to make sure that any smoke that does happen when the board is plugged in won't be accompanied by sparks, tripped circuit breakers, and blown workstations.

"Shorts and Opens" Test

It's always a good idea to test a newly assembled board for shorts between power and ground. Using a good ohmmeter, check the resistance between +5V and ground at a few points. If the board is stuffed then the resistance won't be infinite, due to leakage paths through components. It may vary from several megohms to only a few Kohms. It will also be polarity dependent, so reverse the probes, repeat the measurement, and don't be surprised if the values found are very different.

Check for shorts between ±12V and ground, too, even if your design doesn't use either or both. There might still be an error in the board, after all. Check between +12V and +5V, -12V and +5V, +12V and -12V, and so on, making sure that in each case the resistances you measure match your expectations.

If you do find a short, or a resistance that seems abnormally low, try to obtain a bare PC board and check it. If the problem is there, too, it may be an error in the design itself; check the net-list (a list of all connections on the circuit board) or schematics again, very carefully. It might also be an error in the board fabrication. This could result from mis-registration (misalignment of the board's layers), drilling holes that are off-center or too big, inadequate clearances, poor etching, and so on. Most printed circuit boards are tested at the printed circuit board manufacturer where they are built, usually against one board that is thought to be error-free. Usually this strategy works pretty well, but if the "known good" board really wasn't, then this testing will only prove that all the boards have the same errors.

Fortunately, board fabrication errors are rare. It is much more likely that the short circuit was introduced when the components were installed. Again, test another board if one has been assembled. If both boards have the same problem then look for an error in the bill of materials and the assembly drawings, or look for a part that has been installed incorrectly. If the short circuit isn't evident in both boards, then it is likely to be a workmanship problem. Inspect the board visually again, or use a micro-ohmmeter to track it down (the resistance will be smallest when closest to the short circuit). The latter technique can be tricky, because fractions of an ohm are significant. In skilled hands, though, this type of tool can be very useful.

How to Measure Current Consumption

It's a good idea to measure the current that an SBus card or motherboard consumes, to compare results with those predicted. There are a number of different ways to do this, of course.

If a bench-top type power supply is available, it might be used to first apply power to the card. Often, these power supplies have mechanisms to limit their current output to pre-set levels. This can be used to prevent some catastrophic failure, because the current limit should be set to something just above the typical value expected. If the board tries to draw more than this, the power supply would prevent it. The advantages to this approach are that it is easy and safe. It does not place a workstation or other piece of equipment at risk, either, and current measurements are easily made using an ammeter (either hand-held, or on the power supply's front panel). It provides a good opportunity to test whether any component gets abnormally hot, whether the oscillators (if any) are functioning, and so on. The biggest disadvantage is that it tests the board under abnormal, inactive conditions. This means that any measurements made will probably be low.

Ultimately, the best time to measure current consumption is in place, under test. This is probably also when it is most difficult, unfortunately. Often, some kind of extender board or cable will need to be interposed between the board (an SBus card or motherboard) and its supply (the motherboard or power supply, respectively). Then, the power leads will need to be separated out from the signals. One alternative is to break the connection here and insert a standard ammeter. Another choice is to use a current-loop sensor. This is a device that clips or wraps around a conductor and measures the current in it by sensing the strength of the resulting magnetic fields. These are very useful, because they don't require that the connection be broken and they don't introduce any losses in the line (standard ammeters are never perfectly lossless). Current-loop sensing probes are available for oscilloscopes, too. These can graphically show the current demand patterns of the device under test, and provide feedback on whether or not the board's decoupling is adequate.

Regardless of how the board's current consumption is measured, remember that the results are not worst-case; they are based on too small a sample. Measurement is no substitute for calculations done up-front in the design process.

8.3.3 Test Plan

Surprisingly, a test plan is one of the things most commonly missing when it comes time to debug a new product. All too often, the only plan is to "plug it in and see what happens." This often means that the first few hours or days are spent mulling over the product until some ad hoc plan is developed or falls into place. Do you have diagnostics for the product? Do they provide a thorough test, or are they cursory? If a diagnostic fails, is it a fault in the product, or in the diagnostic? Do you have complete sets of schematics and documentation, including bit locations, definitions, and so on? Do you know what, if anything, needs to be initialized, and how to do it?

It is better to consider all these questions, and many others, during the (relative) lull that normally precedes receiving the first prototypes. In the rush that follows it is all too easy to accept compromises and incomplete solutions in the interest of expediency.

A pre-conceived test plan helps beyond the first prototypes, too; it may also ease production testing. This is because the thought that goes into them may have positive impacts on the design itself. At every turn, and with every feature, the designer should ask, "how am I going to test this?" Can the parts be probed with the board installed? What will you use to trigger the oscilloscope or logic analyzer? Ask also about the exception conditions; unusual operating modes or errors. How will these be generated (or simulated) and tested?

Often, only minor design changes and features can greatly improve the ease with which it is tested. For example, loop-back modes can be included, or spare bits in registers might be used to force certain types of errors. Chip-enables, unused address lines, and so on should never be tied directly to +5V or ground, either; use pull-ups or pull-downs so that the signal can be over-driven if necessary. No matter what the product, ground "stake" pins should be distributed and clearly marked in the silk-screen. This simplifies attaching probe ground leads.

8.3.4 Tools and Supplies

It is also helpful to gather the tools and supplies you will need *before* you need them. There's no substitute for having the right tool for the job, and there is little that is more frustrating than a multi-thousand dollar project stalled for lack of a $1.50 test-clip, EPROM, or other part. Even when you have what you need, it's still no fun to have to scrounge around looking for it. It is better to develop a list of what will be needed beforehand, and gather it together if possible. This activity is a good thing to include in any test plan.

One of the most basic requirements is a good place to work. Preferably, this is a dedicated work area or lab where there is room to spread the equipment out and get comfortable. There should be plenty of electrical outlets nearby, and some provision for getting monitors, cables, and so on out of the way is useful. Anti-static precautions should be taken, too. The table or bench's surface should be conductive, and grounded. Some wrist straps and plugs to ground them are nice, too.

If you are debugging an SBus card, you will need some kind of workstation or other SBus host, of course. If it is the host you are working on, a variety of SBus cards to plug into it are useful. Is your design an interface to some auxiliary component, such as a disk or tape drive, a CRT, a VME bus, or anything else? If so, then make sure to have such a device available, as well as any cables, power supplies, or other components which may be required.

Other test equipment that is useful includes a high-speed oscilloscope; and, if possible, a logic analyzer. Make sure to have enough probes, too; these should be high impedance, low capacitance varieties. By the time this book is published, some test equipment dedicated specifically to the SBus should be available. This includes a timing and protocol verifier card, and another card which captures and displays bus activity. The author has no direct experience with these products, but has used the VME counterparts extensively and found them very useful.

A complete collection of pliers, cutters, screwdrivers, and so on is recommended, as is a fine-tipped soldering iron. Does your design use surface-mount components? If so, special tools may be need to work with them. Do you have socketed PLCC (or similar) components? These may require an extractor tool. Also useful are a grab-bag collection of spare components, including resistors, capacitors, various ICs, and so on. Right-angle "extender" boards are invaluable for probing surface-mount components on the solder side of the board. One example, manufactured by DawnVME Products, is shown in Figure 8.3. Clip leads and test clips are very useful, too, when trying to jury-rig a temporary connection or pull-up, for example.

For FCode development, you will need a copy of the tokenizer program (available from Sun Microsystems or in the *SBus Developers Kits*), as well as access to a machine on which it will run. An EPROM programmer and eraser will be needed too.

8.4 Troubleshooting Clues

It's rare that a new design works perfectly as soon as it is tried for the first time. More than likely there will be a few logic or wiring errors that must be isolated and fixed. If lucky, these bugs will result in fairly obvious and repeatable symptoms. This makes them easy to define and trace. The situation is worse when the symptoms are subtle, seem unrelated to each other, and occur only infrequently. Tracing the problem can be problematic because it is not obvious where to begin, and the very act of attaching probes to the circuit may change its behavior. Experience can be the best guide in these cases.

FIGURE 8.3. *Right-Angle Extender Boards Allow Access To Both Sides of the SBus Card.*

This section summarizes a number of problems that occur frequently when attempting to debug an SBus card for the first time. The material here is distilled from the author's experiences both with his own designs, and with other designs he has helped to debug. After a brief description of the symptom, some possible causes are listed and in some cases, possible solutions as well.

8.4.1 Your SBus-Based Machine Seems to "Hang"

If your workstation ever seems to just stop dead in its tracks, you might consider examining the SBus' **ACK*** lines. The chances are good that you will find all three stuck in the high (de-asserted) state. This is an "idle" acknowledgment code. If you look even closer, you might see that every 256 clocks or so, the level on one or more of them drops just slightly, as if a weak driver were vainly trying to overdrive the signal. In fact, this may be exactly what is happening.

SBus slaves are allowed to drive an idle code onto the **ACK*** lines on any clock after its **SEL*** is asserted in conjunction with **AS***. At some later point it is expected to change the acknowledge code into one that will actually end the cycle. In any case the slave is not allowed to drive the **ACK** lines beyond the cycle's 255th clock. If the card is not working properly, though, it may never drive an appropriate acknowledgment and it might not relinquish the **ACK** lines; it may attempt to drive the idle code ad-infinitum.

On the 256th clock the SBus controller will come along and try to end the cycle, which has timed out, with an error acknowledgment. The SBus controller's drivers may be weak, though, and if weaker than the slave's, this may not work. The end result is that the current SBus cycle will hang indefinitely, and this will stall the whole machine.

Before you do anything else, examine the **SEL*** signals on each SBus slot to determine the offending party. Then, either reset the machine or (if you are daring) use a logic-pulser to overdrive one of the acknowledge lines and get things going again.

The SBus Specification forbids a slave to drive the **ACK*** lines beyond the 255th clock cycle for just this reason. However, this requirement is often ignored. Even when due diligence has been paid in trying to satisfy this requirement, there may still be a problem. If the board is broken enough to time-out in the first place, it may also be too brain-dead to remove itself from the bus at the right time.

Alas, there are three other conditions that may cause your system to hang. All of these conditions are discussed in more detail in Chapter 5. The first is a hardware bug in SPARCstation 1 class machines. This bug prevents time-outs from working at all on write cycles. If your card does not acknowledge a write to it then the system may hang even if you are only driving the **ACK** lines at the proper time.

Another possibility is that the machine may not be truly hung, but brain-dead nonetheless. Again the problem involves write operations on SPARCstation 1 class machines. This time, the bug is in some early revisions of the firmware. If an error acknowledge is received on a write cycle, the Open Boot Prom may trash some of the state in its stack and spin off into Never-Never land.

The last known possibility again involves SPARCstation 1 class machines. Error acknowledgments on burst transfers may cause the SBus controller to lose track of the burst's length. In that case it will not terminate the transfer at the appropriate time.

8.4.2 CRT Stays Blank at Power-up

There are a number of reasons why the screen may stay blank at power-up. The first and most serious is a power supply fault. Check to make sure that the LEDs on the keyboard are flashing properly, and that the normal "beep" tone can be heard. If any of this does not happen then immediately shut the machine down! Check your board for power-to-ground short circuits, or for components that are plugged in the wrong way. Also make sure that the machine is plugged in, and that the wall outlet used has power.

If the keyboard LEDs do flash normally and the "beep" is heard, make sure you wait long enough. The machine may require 8 or more seconds after the LEDs stop flashing before the CRT is enabled. Take advantage of this time to make sure that the CRT is plugged in and turned on. Also check the CRT's cable, making sure that it is firmly seated both in the back of the monitor and in the frame-buffer. Do you have a frame-buffer?

Is the host a SPARCstation 2 (or equivalent)? Is the Open Boot Prom revision 2.0 or higher? If so, the problem may be with reading your ID PROM. If such a machine's firmware probes your board and a time-out doesn't occur, then it expects to find a valid ID PROM. If it does not, then an indefinitely blank CRT may result. Switch to a SPARCstation 1/1+ (or equivalent) machine, or one with a revision 1.X Open Boot Prom. These machines will allow you to get into the Open Boot Prom's debug environment even if the ID PROM is missing or has a stuck data or address bit, etc.

Is the host a SPARCstation 1 (or equivalent)? These hosts seem to be particularly sensitive when oscilloscope or logic-analyzer probes are attached to the **ACK(2)*** signal. If you have done this, try removing the probe.

8.4.3 "Memory Address Unaligned" Errors Occur

If there are "Memory Address Unaligned" errors occurring, it almost always means that a bad virtual address has made its way to the MMU. This can indicate an error in the DVMA master, but it more commonly means that someone else is interfering with the master's efforts to drive the virtual address onto the data lines. Usually, some board is driving the data bus when it shouldn't be.

Check to make sure that your board's data drivers are only enabled at the proper times. **SEL*** must be synchronously qualified with **AS*** before either can be used to enable the data drivers (for more information see page 148). Also, make sure that your card's state machines aren't stalling mid cycle, with the data buffer enables active. If they do, a time-out will usually occur, but the machine will not be able to proceed because the SBus' data lines are being held hostage.

8.4.4 DVMA Errors are Reported

If you receive a large number of DVMA errors reported by the OS while booting or afterwards, one likely cause is that the available SBus bandwidth is being overwhelmed. This is especially true if the DVMA errors are associated with an ethernet device, which can be very sensitive to latency.

One reason this may be occurring is that there might be devices with slow access times which are being accessed frequently. If the problem seems worse with slower SBus clock rates then that is consistent with this scenario. Byte and half-word devices can also throw away significant percentages of the available bandwidth if accessed often. Another possible cause is that some device may be timing-out frequently (which they should not do during normal operation), or is unfairly stealing bandwidth by mis-using atomic operations (see section 5.3.4 on page 170).

DVMA errors can also be reported if the SBus is somehow corrupted by a device. SBus virtual addresses are multiplexed onto the data lines very early in a transfer. If the devices involved in the previous transfer have not quite relinquished control and disabled all their drivers, then it is possible for them to overdrive and disrupt the next cycle's virtual address. In this case, the errors that occur are equally likely to be associated with ethernet, SCSI, or other I/O traffic, whichever is occurring at the time. If you see errors like "le0 memory error," this is the likely cause.

When overlaps such as this occur, it is often because **AS*** is not being used to qualify **SEL***, or that the result is not sampled synchronously. This is discussed in more detail in the section on page 148.

8.4.5 Dynamic Bus Sizing Does Not Seem to Occur

If the **Siz** code on any (non-burst) transfer exceeds the slave's port width then bus-sizing should occur. This means that the master will break the transfer up into either two or four smaller transfers.

If this does not seem to be happening, then first determine if the master generating the transfer supports bus sizing. Most current masters do, the B.0 revision of the specification strongly recommends it, and future revisions of the specification may require it.

If the master does support bus sizing, then it is highly likely that the master does attempt to perform follow-on cycles, as required, but the slave does recognize that one transfer has ended and another has begun. More information on this can be found in Section 8.4.8, and in a related discussion starting on page 166. In brief though, the scenario is this: after the controller samples the **Ack*** signals and finds them asserted, it immediately de-asserts **AS***. After only one clock cycle, however, **AS*** is asserted again. This occurs so fast and so soon that the slave may not recognize it. After all, the slave is still driving data (if a read) and it is actively driving the **Ack*** signals off. If the slave's state machines do not also sample **AS*** during this clock they will stall, waiting for the end of the transfer that has already come and gone. A time-out will be the inevitable result.

8.4.6 Unexpected (or Phantom) Interrupts

It's possible to get error messages associated with interrupts, even if your card is not generating interrupts. This has been particularly maddening to developers, who have occasionally gone as far as to cut the traces on their card attached to the interrupt lines, only to find that the phantom interrupts are still there. Usually, the error message looks something like the following:

```
Booting from: sd(0,0,0)vmunix
Level 15 Interrupt
ok ->
```

Unfortunately it has not yet been possible to find a conclusive cause-and-effect relationship with these interrupts.

One piece of information that might help track down the source of this problem is that on some SBus hosts, write errors are handled differently than read errors. Read errors are treated like synchronous bus errors, but write errors are considered asynchronous and are handled using level 15 interrupts. The operating system may hide this level of detail from the user, but the Open Boot PROM does not.

These phantom interrupts also seem particularly common when using LSI LoGIC's L64853 or L63853A SBus Interface Chip. In at least one case the cause was tracked down to the developer's mistaken assumption that this chip's internal ID register could be used instead of an ID PROM. This is not the case. The internal ID register is the historical remnant from very early revisions of the SBus Specification, and is only useful now during chip testing. When a valid ID PROM was added to this developer's board, the phantom interrupts went away.

The lack of a valid ID PROM has been an implicating factor in many cases of phantom interrupts. Another time, the problem was traced to a board asserting error acknowledgments at power-up, and at other inappropriate times.

One other possible cause of phantom interrupts is the relatively long rise-time of the interrupt signals themselves. It is possible for the processor to mistakenly recognize an interrupt when in fact the signal has been de-asserted, but is rising slowly and has not yet passed the threshold. For more information please see a related discussion which begins on page 153.

8.4.7 The Card's FCode Is Not Recognized During SBus-Probing

A card's ID PROM must be able to respond to both 8-bit and 32-bit **SIZ** codes. SBus systems may start probing with either request size, and if the slave does not support one or the other, the system might mistake the resulting error acknowledgment for a time-out. As a result, it would subsequently ignore that slot and not recognize its FCode.

If you cannot find your device in the host's device tree, or if its FCode words aren't found in the OBP Toolkit's dictionary space after probing, then it is possible that the board is not allowing both 8- and 32-bit accesses. There is no reason for the slave to intentionally restrict accesses, either, because 8-, 16-, and 32-bit access capabilities do not complicate the design of a slave, and may even simplify it (see the discussions related to dynamic bus sizing elsewhere in this book).

A card's FCode might not be recognized if the slot that it is plugged into is not probed. The OBP Toolkit's **sbus-probe-list** environment parameter controls this. (See section 4.2.1 on page 116.)

8.4.8 Data Access Errors Occur Without an Error Acknowledgment

In some cases it is possible for a *Data Access Exception* to be reported even if the slave seems to be generating a valid (non-error) acknowledgment. One possible cause is the timing of the

ACK signals. If the required setup or hold times are not met then they may not be correctly sampled. The acknowledgment received might be very different than that intended. Accidental assertion of the **LateError*** signal is another possible cause.

If bus sizing should be taking place, and the master is part of the host in a host-based system, then there is another possibility to investigate. Host masters in host-based systems are not required to perform the translation phase of an SBus transfer. The follow-on cycles necessary to complete a bus-sized operation may occur very quickly; **AS*** may be de-asserted for a single cycle exactly after the **ACK** signals have been sampled. During this clock the slave's state machines should be actively de-asserting the **ACK** signals due to the requirements of *active-drive*. Then they normally proceed to a state where they wait for **AS*** to be de-asserted, indicating the end of the transfer. But **AS*** has already been re-asserted for the follow-on cycle at this time! The slave will probably mistake the follow-on cycle for the last part of the first transfer. It will not respond with another **ACK**, and the follow-on cycle will then time out. Unless looking for this kind of problem it can be difficult to find. If using an oscilloscope, you may not be able to pick out the single cycle de-assertion easily. All else will look almost exactly as if a single transfer is occurring, with proper acknowledgment, but with an error nonetheless.

8.4.9 Apparent Slot Dependencies

A properly designed SBus card should be slot-independent. It should not matter which slot in a machine the card is plugged into. Still, design flaws or misunderstandings can cause problems which manifest themselves as apparent slot dependencies. For this reason it is prudent to try your SBus card in a variety of slots, looking for differences in behavior.

FCode Problems

Word definitions like **my-address** allow a card's FCode to be designed in a slot-independent fashion. It's possible to inadvertently hard-wire in constants or values which will result in slot-dependencies.

It is also possible that the success or failure of a device's FCode may depend on the order in which it is probed. For example, suppose that a card needs to map a large address space, as do other SBus cards in neighboring slots. Toolkit mapping resources are limited during probing, and conflicts may arise. If the card being tested is probed early then it might have no trouble mapping all the address space that it needs. If the card is probed later, there may not be enough mapping resources left to satisfy its needs. As the order in which slots are probed is slot-dependent (see the description of the **sbus-probe-list** parameter in section 4.2.1 on page 116) the resulting problems may appear to be, also.

Slave-Only Slots

Some SBus hosts and expansion boxes have slots which are meant only for slaves. Slot #3 in the SPARCstation 1 and 1+ machines is an example. If the slave interface on your board works but the master interface doesn't, this is one thing to check.

This is also something to be careful of because the pins dedicated to **BR*** and **BG*** will be reserved, and their value is not guaranteed. It is possible that the master interface might wrongly believe it is given access to the bus. If this happens it may disrupt the entire machine's operation. It is even possible that damage will occur because of the bus-contention that results.

Timing Problems

Some of the most likely causes of apparent slot dependencies are problems with signal timing. The author has recent experience with one such case, where the problem was eventually tracked down to **ACK*** signals that were asserted too late in the clock cycle. The board involved was being debugged in a 25 MHz SBus machine (where timing will be most critical), and the timing was such that the board just worked in slot 2, but would not work consistently in slot 1.

Once found, the solution to this kind of problem is to bring the timing within the specified requirements. Choose a faster driver, or speed-select your parts. Reduce trace lengths or signal loadings or logic levels. Take a close look at your clock distribution and eliminate as much delay and skew as possible.

Finding the problem, though, can be much easier said than done in many cases. The best approach is the proverbial "ounce of prevention." Careful timing analysis during design can save many frustrating hours in the lab. Also, any problems that are found are usually much easier to fix during design, because once a design has been committed to fiberglass and silicon there are far fewer options available.

No design analysis will ever be perfect, though, and some problems may get through. Fortunately, if you suspect timing problems in your design there are still a number of things you can do.

Try a different slot as mentioned above, or a different machine (preferably one with a slower or faster SBus clock). Consider heating and/or cooling the parts involved, because timing can be temperature dependent. In all cases carefully note any differences in behavior you find, and use them as clues to help deduce what might be happening. If there are extraneous boards in the machine then pull them out if you can; their capacitance will be slowing signals and there may be other interactions. Get a good scope or a logic analyzer with resolutions of 5 nanoseconds or less (make sure to de-skew the probes), and examine the signals being driven by your board. Also keep in mind that if your timing is right on the edge, a scope probe may change it! This can be frustrating—just when you think you're getting close the symptoms change or disappear! It's a very good clue, however, that you really are on the right track.

Electrical Noise Problems

Electrical noise problems can often manifest themselves as timing problems, and in fact the two are often closely related. Electrical noise can increase the time required for a signal to settle to its final value, effectively increasing the propagation delay. For a detailed discussion of what this noise is and how it is generated, see Chapter 7.

Noise problems can be even harder to identify than timing problems. Some amount of noise is always present, and determining what is normal and what isn't can be very tricky. Also, examining noise with a scope is difficult to do. The scope probe itself will change the signal you are observing, and even with the very best of scopes the display is only an approximation of what is actually present. (This is due to non-linearities, and limited frequency responses, DC and AC offsets in the scope ground reference, changes due to added capacitance, inductance, impedance, and so on.)

Here, prevention really is the best strategy. Detailed information on the causes of noise how to reduce its effects is contained in Chapter 7.

8.4.10 Bad WRITES

If it appears that write operations seem to be corrupting data, a likely cause may be hold time violations. The slave must latch the data on the clock edge *before* the one in which the acknowledgment is sampled by the master. This means that it must de-assert any on-board write strobes before it drives the **Ack*** lines; write data is not guaranteed to be valid *during* the clock in which the acknowledgment is driven. Failure to understand this and design the slave properly will most likely result in hold-time variations and corrupted data. A more detailed discussion on this matter can be found starting on page 156.

Acknowledgments

SPARC and MicroSPARC are registered trademarks of SPARC International, Inc.

References

Ott, H. W. *Noise Reduction Techniques In Electronic Systems*. John Wiley and Sons, 1976.

SPARC International, *SPARC-Line (Newsletter)*, September 1991.

SERFboard User's Guide

9

9.1 Description

The SERFboard is a general-purpose SBus slave interface card, intended for use as a prototype development platform. It is capable of responding to 8-, 16- or 32-bit SBus transfers, and can also perform burst operations. The SERFboard buffers virtually all SBus signals and performs the necessary handshaking, freeing the user to concentrate on their added hardware.

A block diagram of the SERFboard is shown in Figure 9.1. The board is divided up into two major parts. The first part is the built-in interface circuitry. This is composed of the SBus interface itself, and a "prototype" interface which is isolated from the SBus and designed to allow the simple handshaking with whatever prototype circuitry the user wishes to attach. The distinction between SBus interface and prototype interface is largely a logical one, because in reality, the state machines of each are closely connected.

The second major part of the SERFboard is the integral prototype grid-work capable of holding up to three 40-pin DIP integrated circuits. If this area is not sufficient, additional prototype area can be added via a variety of optional cards that mount either perpendicular or parallel to the SERFboard. The SERFboard and these expansion options are shown in Figure 9.2.

The SERFboard's interface circuitry is controlled almost entirely by state-machines in three 22CV10-type PALs. The source files for the PALs used are included in this documentation. While the existing programs should be sufficient for most applications, it is possible to customize the PAL state machines for a particular application, eliminating the additional "glue" logic that might otherwise be needed. More information on customizing the PALs to your application can be found in Section 9.7 below.

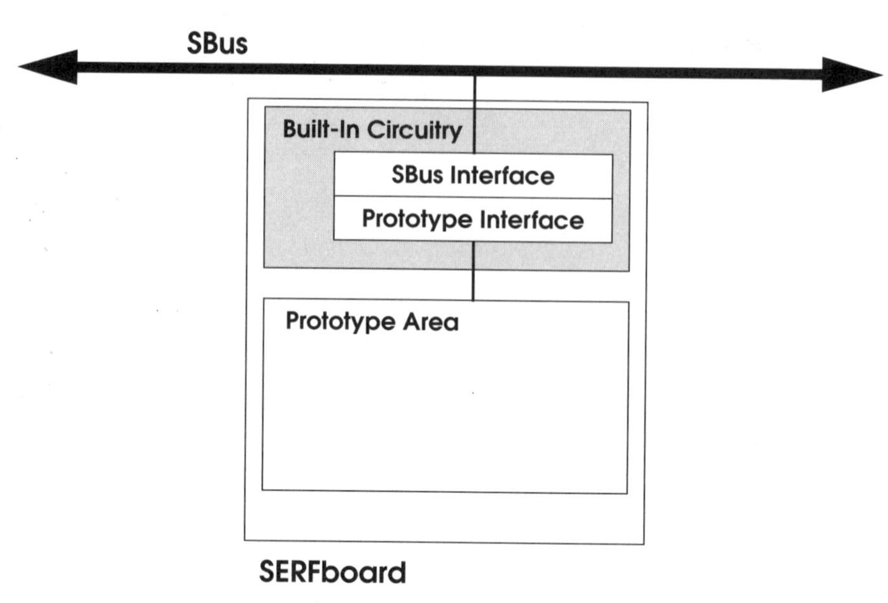

FIGURE 9.1. *SERFBoard Block Diagram.*

9.2 Programmer's Model

This section describes the SERFboard's interface as seen from a programmer's (or driver's) perspective. In and of itself, the SERF-board has no registers or other on-board devices. It serves only as a conduit between the SBus and whatever logic is added by the end-user.

9.2.1 Address Spaces

The SERFboard only uses 25 of the 28 SBus physical address bits. This is to ensure compatibility with SBus hosts which only provide the 25-bit physical address subset.

The SERFboard is designed to quickly and easily accommodate prototype designs that may vary in width or access times. To simplify this task without requiring jumpers or DIP-switches, the 25-bit address space of the board is divided up into four sections: one for the PROM; and one each for 8-bit, 16-bit, and 32-bit wide devices. These four spaces are then further divided into sub-

spaces which control the access timing. Figure 9.3. illustrates
these divisions.

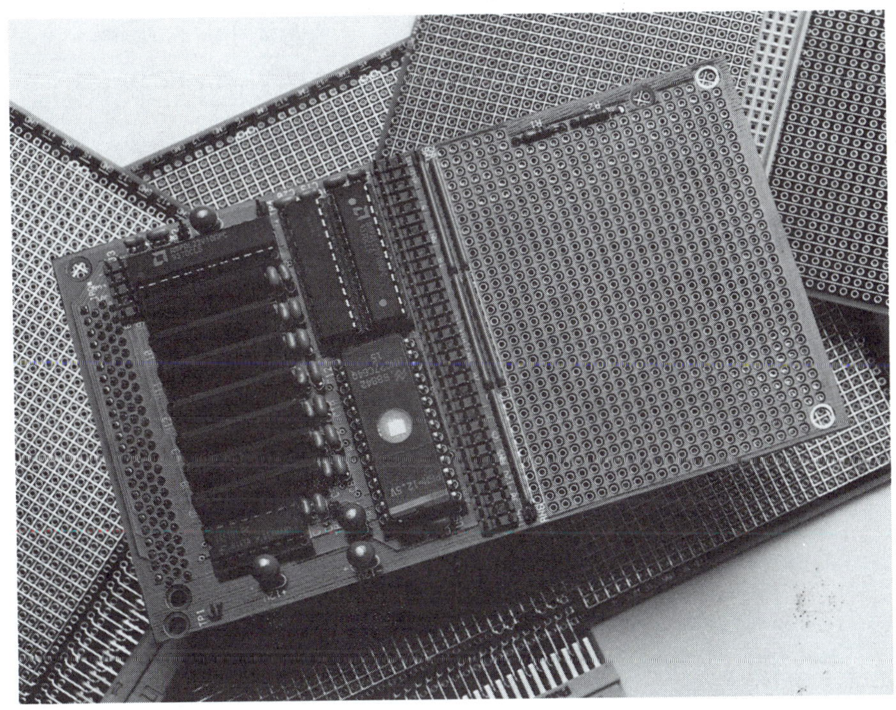

FIGURE 9.2. *SERFboard and Expansion Options.*

The number of "wait-states" refers to the number of additional
states inserted in any access before the acknowledge is generated.
This is covered in more detail in Section 9.4.2 below.

9.3 SBus Interface

This section discusses the characteristics of the SERFboard's SBus
interface. All SBus signal lines (except **LateError***) are buffered.
This provides protection and isolation both to the SBus and to logic
added to the SERFboard.

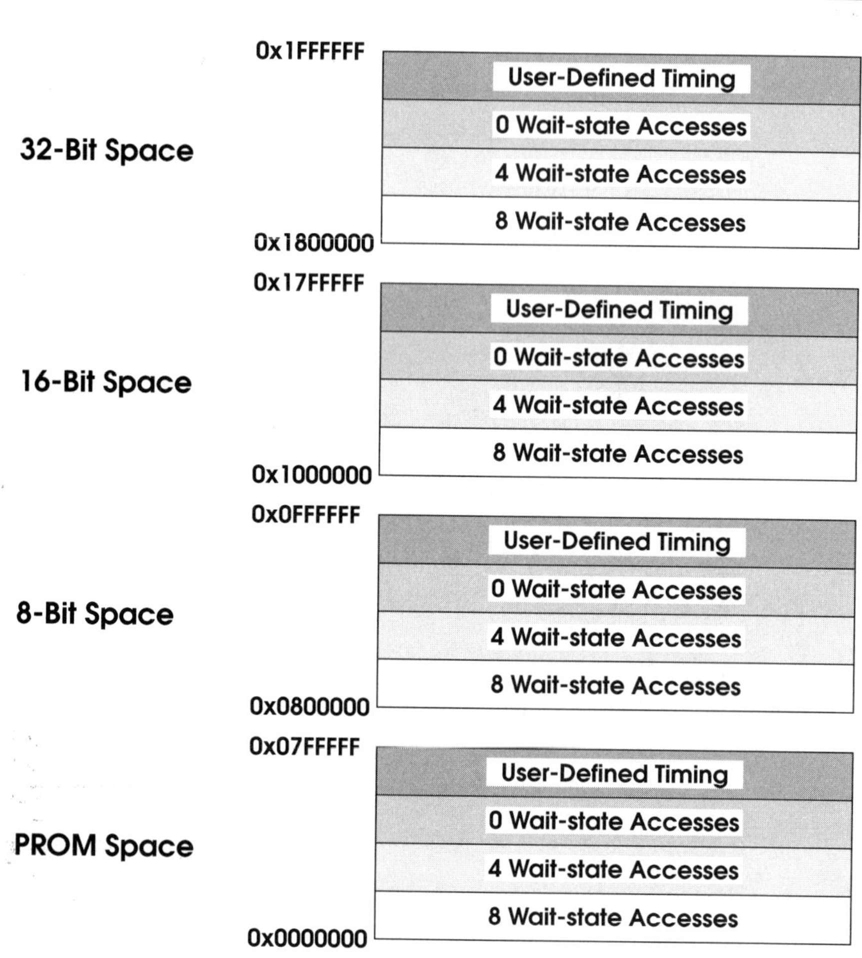

FIGURE 9.3. *SERFboard Address Map.*

9.3.1 SBus Compliance

A rigorous effort has been made to comply fully with SBus Specification B.0. There are certain discrepancies, however:

- Stub lengths for some of the signals from the SBus connector exceeds the 2" maximum specified in Revision B.0. The **Ack*** signals are among the worst, at approximately 3.5" (89 mm).

- Address-wrapping on burst transfers will not work properly for anything but 64-byte bursts without additional logic. Burst transfers do work in a specification compliant manner if burst starting addresses are aligned properly. See Section 9.3.3 for more information.

9.3.2 Electrical Characteristics

The DC electrical characteristics of the SERFboard are summarized in Figure 9.4. The values represent the worst case values for any of the SBus signal lines. Usually this means that the specification sheets for the PALs and the 74FCT245/244 buffers were examined, and the worst-case value was chosen as representative.

Symbol	Parameter	Test Conditions	Min	Typ	Max	Units
Cio	I/O Capacitance	Vio=0V	-	16	20	pF
Iin	Input Leakage Current	-	-	-	10	uA
Iout	Output Leakage Current	-	-	-	10	uA
Iack	Ack* Leakage Current	-	-	-	10	uA
Vih	Input HIGH Voltage	-	2.0	-	-	V
Vil	Input LOW Voltage	-	-	-	0.8	V
Voh	Output HIGH Voltage	Iol=12 mA	-	-	0.5	V
Vol	Output Low Voltage	Ioh=3.2 mA	2.4	-	-	V

FIGURE 9.4. *DC Electrical Characteristics.*

9.3.3 Transfers and Modes Supported

The SERFboard is capable of performing byte, half-word, and word transfers. It will also perform 16-word bursts, and in some cases 2-, 4-, and 8-word bursts.

In order to support burst transfers, there is a built in address counter that will increment **PA(5:2)**. **PA(1:0)** must both be low for word addresses, which are required for all burst transfers. For simplicity and cost savings full wraparound addressing was implemented only for 16-word bursts. Bursts of other sizes may still be performed, but will only work properly if the address is aligned to the burst size boundaries. For example, **PA(2:0)** must be low for 2-word bursts, **PA(3:0)** for 4-word bursts, and **PA(4:0)** for 8-word bursts.

While not strictly specification compliant, this restriction will be of little consequence in most applications where burst transfers need to be made to SBus slaves. If full compliance and address-wraparound is needed for all burst sizes, this may be accomplished with PAL modifications and additional hardware. Please contact Dawn VME Products for more information.

9.3.4 Transfers and Modes Not Supported

The SERFboard will not perform 64-bit wide SBus transactions and will return an error acknowledgment if these are attempted.

The SERFboard does not generate or check the SBus' optional **DATA** parity signal. Also, there is no standard support for the generation of rerun acknowledgments.

9.3.5 FCode EPROM

A 28-pin, 0.6 inch wide socket is provided for the FCode EPROM. This socket accommodates standard 32k x 8 CMOS EPROMs. For proper operation, the access time of any part used should be 250 nanoseconds or less. (See Section 9.4.5 for more timing information.)

Note that the physical orientation of the EPROM is opposite to that of the nearby PAL devices. This is due to board routing constraints. Care should be taken to guarantee pin 1 of the EPROM always lines up with pin 1 of the socket.

9.4 Prototype Interface

This section discusses the *prototype* interface which the SERFboard provides for attachment of prototype logic. While the SBus interface is synchronous, the SERFboard's prototype interface assumes that most logic the user adds will be asynchronous. This allows for easy use of microprocessor-compatible peripherals. It also allows added circuitry to generate and use its own clocks, which may vary in frequency from that of the SBus. Or the SBus clock may be used, but with less concern over skew, setup, and hold times.

9.4.1 Signal Summary

Figure 9.5. on page 302 shows the pin-out for connector J2. This connector contains the signals needed to interface a prototype design to the Prototype interface on the SERFboard.

Wires may be attached directly to the pins on these connectors. A cable may also be attached and routed through the backplate to an external prototype. Alternatively, another piggyback board with a larger prototype area may be attached via these connectors.

PAL Clock

The SBus **Clock** signal is buffered by the SERFboard, then routed to each of the PAL devices. It is also routed past the prototype grid area for use by any logic added there. It is important to realize that because this signal passes through one level of buffering, it will have added delay (1.5 - 6.0 ns) with reference to the actual SBus clock.

It is important to avoid simply attaching a wire somewhere onto this net. It is routed very carefully using transmission line techniques, to optimize the signal quality at all points along the way. The trace runs sequentially from the source, through each load, and finally to the parallel termination network formed by resistors R1 and R2. There are no stubs or branches along the way, and this characteristic must be preserved when adding a load to this signal. Otherwise signal reflections and ringing may occur, which could lead to excessive delay or false triggering.

An example of the wrong way to add a new load is shown in Figure 9.6. Here, the new load has been tied directly to one of the original loads, usually with a piece of wire. While this is the simplest method of adding a load, the resulting net is no longer sequential. There is now a branch encountered by any wavefront traveling from the driver. One fork contains the remainder of the original net and the terminator, the other fork contains the new load. If the wire used is long it forms an unterminated stub. If the wire is short then the two loads encountered appear to be one, larger lumped load. In either case the result is an impedance discontinuity and perhaps reflections and ringing.

PIN	ROW A	ROW B	ROW C
01	GND	N/C	+5V
02	GND	INTA*	+5V
03	GND	BUF_RST	+5V
04	BUF_RST*	SL_08BIT*	GND
05	NOT_USED	SPARE	GND
06	Count*	ERR*	+5V
07	Data(4)	Data(17)	+5V
08	N/C	Data(1)	GND
09	Data(0)	BUF_WR*	BUF_RD*
10	Data(5)	Data(3)	Data(2)
11	Data(7)	SL_16Bit*	RDY*
12	Data(6)	SL_32Bit*	Data(10)
13	Data(14)	Data(13)	Data(11)
14	Data(12)	Data(9)	Data(15)
15	Data(8)	Data(16)	Data(19)
16	Data(22)	Data(21)	Data(23)
17	N/C	N/C	N/C
18	Data(25)	Data(26)	Data(18)
19	Data(29)	N/C	N/C
20	PA(18)	Data(28)	Data(27)
21	Data(24)	PA(3)	Data(30)
22	PA(1)	PA(0)	Data(31)
23	PA(4)	PA(2)	GND
24	PA(14)	PA(10)	+5V
25	PA(15)	Data(20)	+5V
26	PA(5)	PA(17)	GND
27	PA(7)	PA(6)	GND
28	+12V	PA(11)	-12V
29	+12V	PA(12)	-12V
30	+12V	SB_LERR*	-12V
31	PA(8)	PA(9)	PA(16)
32	SB_Data_Parity	PA(19)	PA(13)

FIGURE 9.5. *J2 Connector Pin-out.*

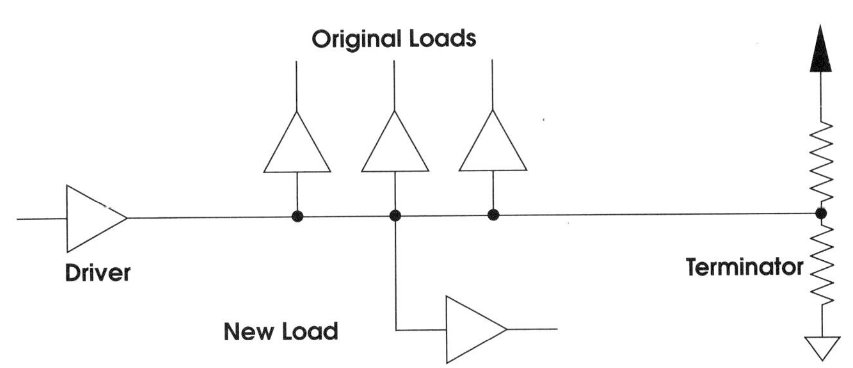

FIGURE 9.6. *Adding a Clock to the PAL Clock Trace (The Wrong Way).*

To properly add a load to this signal, the loads must remain distributed and sequential. The easiest way to accomplish this is to first cut the trace. Then a wire is run from the driver side of the cut trace to the new load or loads (if there is more than one new load they must be daisy-chained). Finally, a wire is run back to the terminator side of the cut trace. An example of how this can be done is shown in Figure 9.7.

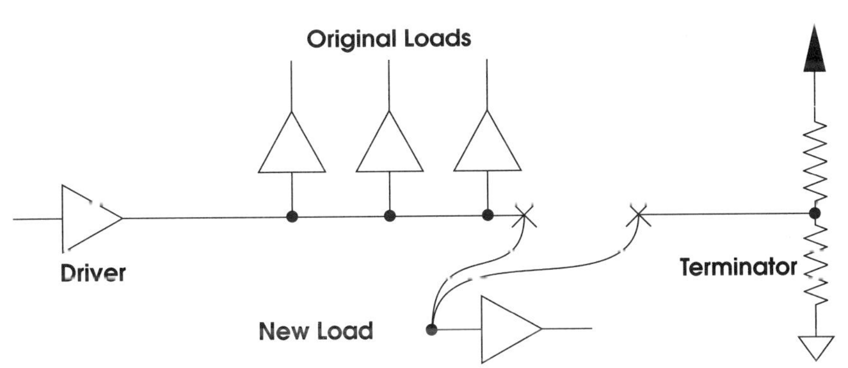

FIGURE 9.7. *Adding a Clock to the PAL Clock Trace (The Correct Way).*

To simplify adding loads to the PAL clock signal, there is a special trace pattern on the bottom (solder-side) surface of the board. The location of this pattern and its structure are shown in Figure 9.8. To connect a new clock load, use a knife or scribe to cut the short trace connecting the two pads. Then daisy-chain a wire from one pad, through all loads, and back to the other pad. Note that if the wire connection is very short (less than 1 inch or so) then it is probably acceptable to connect a single wire from the new load to one of these pads (but nowhere else on the trace). In that case the trace should not be cut.

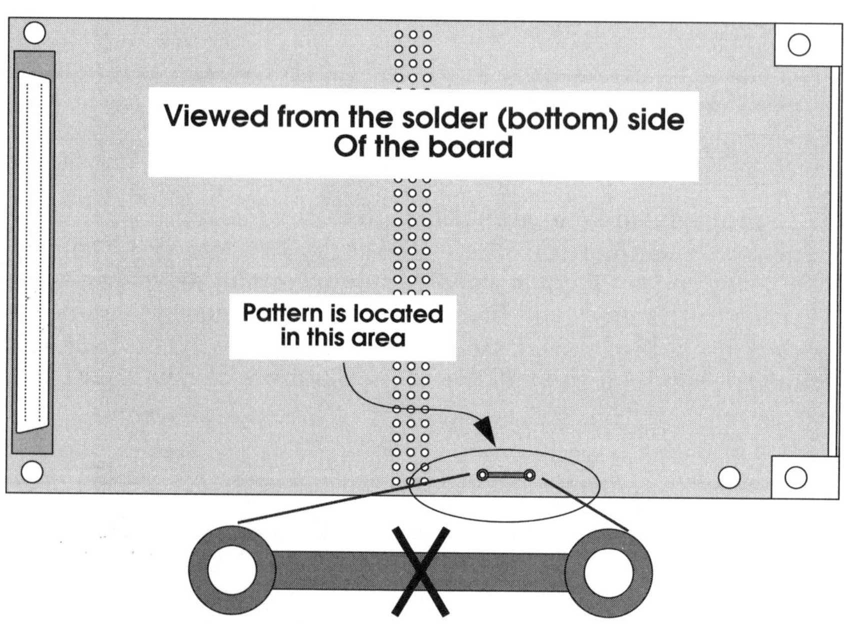

FIGURE 9.8. *PAL Clock Trace Pattern For Adding Loads.*

Data Lines

The **DATA(31:0)** lines make up the SERFboard's internal 32-bit wide data bus. These lines are isolated from the SBus'

sb_data(31:0) lines by transceivers. Devices of varying widths should be connected to the data lines as follows:

* Devices that are 8-bits wide should be connected to **DATA(31:24)**.

* Devices that are 16-bits wide should be connected to **DATA(31:16)**.

* Devices that are 32-bits wide should be connected to **DATA(31:0)**.

Please see Section 9.4.2 below for information on the timing relationships of these signals to the SBus and to the other interface signals.

Address Lines

Of the 25 SBus address bits brought into the SERFboard, the uppermost four (**PA(24:21)**) are decoded as described in Section 9.2.1 above. **PA(19:0)** are buffered, and provide a 1 Mbyte space in each of the four device regions (PROM, 8-bit, 16-bit, and 32-bit). **PA(20)** is not decoded or buffered.

Please see Section 9.4.2 below for information on the timing relationships of these signals to the SBus and to the other interface signals.

Device Space Selects

There are three device space selects generated by the Prototype interface on the SERFboard. **SL_08BIT***, **SL_16BIT***, and **SL_32BIT*** for 8-, 16-, and 32-bit devices, respectively. These lines are driven low when the appropriate device space is selected, and are already synchronously qualified with **AS*** and **SEL*** so that no further qualification is necessary.

On write cycles these selects will de-assert at the end of the cycle before the ACK signals are driven. This guarantees write data hold time. The **BUF_WR*** signal is a level and should not be used as a write strobe. If the circuit being prototyped needs a write strobe then the **BUF_WR*** should be gated with the appropriate **SL_XXBIT*** signal in a glitchless fashion. One possible method of doing this can be found in Figure 9.18. on page 320.

Controls

A number of control signals are used by the interface on the SERFboard:

- The **BUF_RD*** signal indicates the direction of the transfer. This signal is asserted (low) by the interface logic whenever data is to be transferred from the slave (the SERFboard) to the master. Otherwise this signal is de-asserted (high). This signal is a level, not a strobe.

- The **BUF_WR*** signal is an inverted copy of **BUF_RD***.

- The **ERR*** signal is used to signal an error back to the master which initiated the current cycle. This is an input to the interface logic. If asserted (low) by add-on circuitry this will terminate the current cycle with an immediate error acknowledgment. This signal is double-synchronized, and is pulled-up into the de-asserted state by a resistor on the SERFboard so that it may remain unconnected if not needed.

- The **RDY*** signal is used to signal the completion of a cycle that requires a variable number of wait-states. This is an input to the interface logic. When asserted (low) by add-on circuitry this will allow the SBus interface to assert an acknowledgment, ending the transfer. This signal is double-synchronized, and is pulled-up into the de-asserted state by a resistor on the SERFboard. Note that if this signal is unconnected and an access is made to a **RDY*** dependent address space, then the transfer will time out.

- The **INTA*** signal is an input to the Prototype interface which is used to indicate that an interrupt should be generated when asserted (low). Before an interrupt can be generated, a jumper must be installed to connect it to the proper SBus interrupt level. This is discussed in more detail in Section 9.4.4 below. This signal is pulled-up into the de-asserted state by a resistor on the SERFboard so that it may remain unconnected if not needed.

- The **BUF_RST*** signal is a buffered copy of the SBus **RST*** signal. This signal is used to reset the SERFboard's Prototype interface state machines, and may also be used to reset any prototype hardware that gets added. The SBus Specification states that this signal will be asserted for at least 512 bus clock cycles. Timing on the leading edge is not guaranteed (in fact the leading edge itself is not guaranteed), but the trailing edge is supposed to meet setup and hold requirements. See Section 9.4.2 below for more information.

- The **BUF_RST** signal is an inverted copy of **BUF_RST***. It is provided for convenience, and may be connected directly to components that have high-true reset inputs.

- The **SB_LERR*** signal is provided for those applications which must generate (or monitor) **LateError***. This signal is connected directly from the SBus connector to the J3 jumper block. There is no buffering of any kind provided. In most cases, this signal will not be connected and can be ignored.

- The **COUNT*** signal, when asserted, instructs the 74FCT191 address counter to advance the word address. Normally this signal would not be used by prototype logic that the user adds to the SERFboard. It might be used as an indicator of the separation between words in a (multi-word) burst operation.

- The **NOT_USED** and **SPARE** signals are just what the names imply. These signals are an unused output and input (respectively) of the *MISCPAL*. These signals are brought to the J2 connector for ease of use, and the user is free to reprogram the PAL and make use of them in any way necessary. For example, they could be used to buffer, invert, or synchronize a signal. One or the other could be used to bring additional signal terms out of or into the PAL, as well. **SPARE** (the input) is pulled up by a 10K resistor on the board.

9.4.2 Access Timing

Virtually all signals to and from the SBus are buffered by the SERFboard. This buffering is necessary to isolate prototype logic from the bus, keeping trace lengths, leakage currents, and capacitance within limits. Unfortunately, this buffering adds delay to the bus signals. This added delay must be accounted for when analyzing the timing requirements of the logic that the end user adds to the SERFboard. The clock signal is also buffered, as discussed earlier, and its added delay also must be considered.

The buffers and transceivers used on the SERFboard add one to six ns of delay, overall. With respect to the SBus clock (at the SBus connector, not the buffered version) this subtracts 6 ns from the signals' setup times (leaving 9), but adds at least 1 ns to the hold time (adding up to 1, total; 0 is the usual number).

The clock has a similar delay, but this delay and the others are independent; while one is "fast," the others might be "slow," or vice-versa. For example, assume that the physical address or (write) data buffers delay those signals by 6 ns, but the clock buffer delays it by only one. The results is that these signals are delayed $(6 - 1)$ = +5 ns with respect to the buffered clock. Setup times are decreased and hold times are increased by just that amount.

Now assume that the situation is reversed. In this case the delay is $(1 - 5) = -5$ ns. In other words, the physical address and write signals now change state *earlier* with respect to the buffered clock. The setup times are increased, which is helpful, but the hold times are decreased to –5 ns. This value is not enough in many cases.

The prototype interface need not be synchronous, however, for both these reasons and because so many of the devices that may be added to this prototype board are asynchronous. Neither the SBus clock nor the buffered version need be used as a timing reference; the **SEL_XXBIT*** signals provide a reference which is often easier to use. Using that as an assumption, the SERFboard's AC characteristics are summarized in Figure 9.9. The rest of this section discusses timing related issues.

Symbol	Parameter	Conditions	Min	Typ	Max	Units
Tas	Address setup to SL*	rd or wr	10	-	-	ns
Tca	Hold time from CLK	rd or wr	2	-	-	ns
Trs	RD* or WR* setup to SL*	rd or wr	3	-	-	ns
Tcr	Hold time from Clk	rd or wr	2	-	-	ns
Tra	Read access time	0 wait-states	55	-	-	ns
		4 wait-states	215	-	-	ns
		8 wait-states	375	-	-	ns
Taw	Addr stable before SL*	-	10	-	-	ns
Twa	Addr stable after SL*	-	32	-	-	ns
Tdw	Wr Data stable before SL*	-	10	-	-	ns
Twd	Wr Data stable after SL*	-	32	-	-	ns
Tww	SL* pulse width	0 wait-states	35	-	-	ns
	(on writes)	4 wait-states	195	-	-	ns
		8 wait-states	350	-	-	ns

FIGURE 9.9. *AC Switching Characteristics.*

General Cycle

The timing diagram for a read operation is shown in Figure 9.10. The prototype interface is designed to connect easily to most devices. Without the insertion of any wait-states, **SEL_XXBIT*** (any of the **SEL*** signals) is active for two clocks on read cycles. This is approximately 80 ns with a 25 MHz SBus clock. Setup time from the address to the select will vary greatly from one host to the next, but is guaranteed to be at least 10 ns. Setup time from **BUF_RD*** to the select is guaranteed to be at least 3 ns. Hold time for both (from the rising edge of the clock used to sample the data) is at least 2 ns. The accessed device must guarantee that data is available and valid during the clock after the SBus interface generates the acknowledgment, as shown.

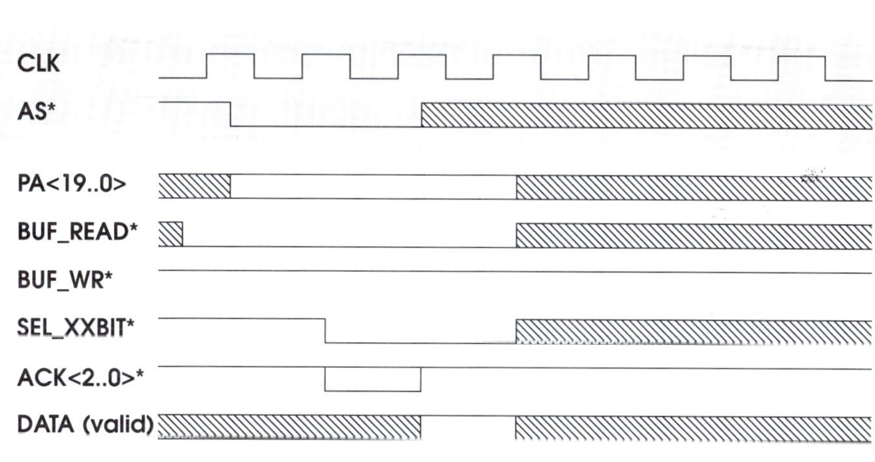

FIGURE 9.10. *Basic Prototype Interface Timing (Read operations).*

The timing diagram for write operations is shown in Figure 9.11. Notice that one additional clock is required for write operations. This is necessary to guarantee data and address hold times (the SBus master provides valid data at the times shown). Otherwise, timing here is much the same as it is in the case of a read operation. The major difference is in the **SL_XXBIT*** signal, which is used as a strobe. For write operations this signal is de-asserted before the cycle in which the **ACK** signals are driven, when address and data are guaranteed to still be stable. With no additional wait-states, the width of **SL_XXBIT*** will be one SBus clock interval. Both data and address are held for at least one SBus clock as well. (1 SBus clock interval equals 40 ns at 25 MHz.)

One consequence of using **SL_XXBIT*** as a strobe is that it must be glitchless. Also, any sub-select decoding (dividing the 8-bit space among four different devices, for example) must also be done in a glitchless way if the strobe properties are to remain. This should not be difficult, however, because the address is stable both before and after **SL_XXBIT*** is asserted.

Strobing write operations with the device space select signals (instead of with a dedicated write strobe) is adequate for most devices which might be connected to a SERFboard. If a write-strobe is needed then the **BUF_WR*** signal can be gated with the appropriate select signal in some glitchless fashion.

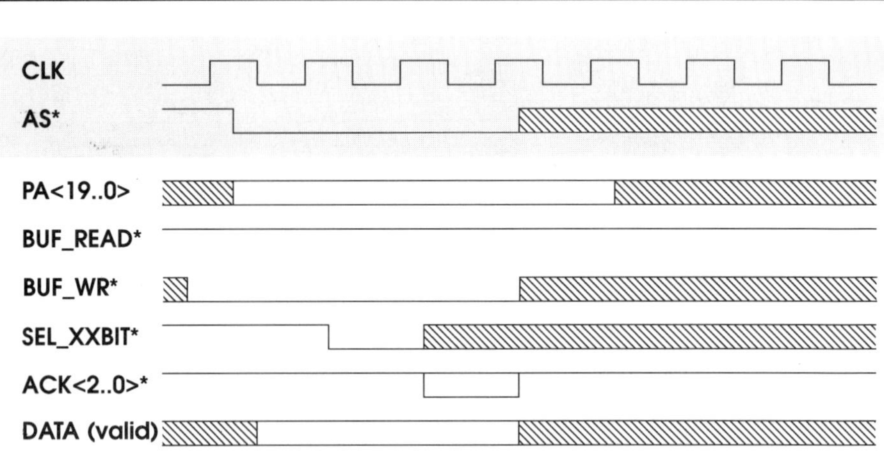

FIGURE 9.11. *Basic Prototype Interface Timing (Write operations).*

If the minimum access times are not sufficient for a particular application, they can be easily stretched out by adding wait-states. Either four or eight wait-states can be added via special address decodes (as discussed in Section 9.2.1 above). An example of either a read or write access with four additional wait-states is shown in Figure 9.12. The wait-states are added in such a way that both read and write cycles are stretched. With 8 wait-states inserted,

375 ns read accesses occur, and write pulses with 355 ns minimum widths are generated. These numbers are not exact multiples of the 40 ns clock interval due to on-board delays and skews.

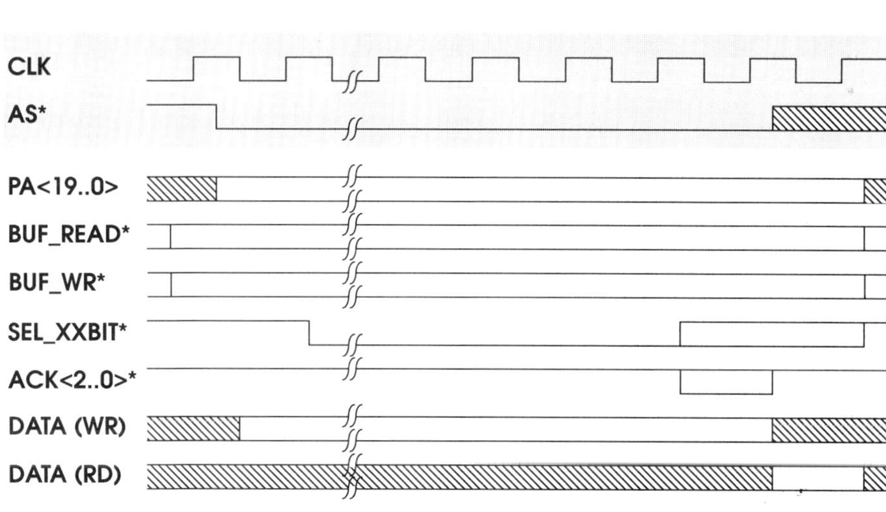

Timing shown includes four wait-states (extra clock cycles).
Additional wait-states may be added where shown.

FIGURE 9.12. *Prototype interface timing with wait-states (Reads or Writes)*

User-Defined Timing

If even eight wait-states are not sufficient, or if the delay needed is variable, then the **RDY*** signal may be used to insert an arbitrary number of wait-states. A timing diagram showing this mechanism is shown in Figure 9.13.

Please remember that the SBus specification allows at most 255 clocks in any access. The SERFboard does not enforce this maximum, and it is the user's responsibility to guarantee that this maximum is not violated.

Terminating a Cycle with an Error

It is possible to force the SBus interface logic on the SERFboard to terminate the current cycle with an error acknowledgment. Generally, such error acknowledgments are issued if the transfer size is illegal or not supported, but it may also be used to indicate some internal error condition has occurred. Slaves should not use error acknowledgments for flow-control purposes, or to signal status. For more information on the uses of error acknowledgments, please see the SBus Specification.

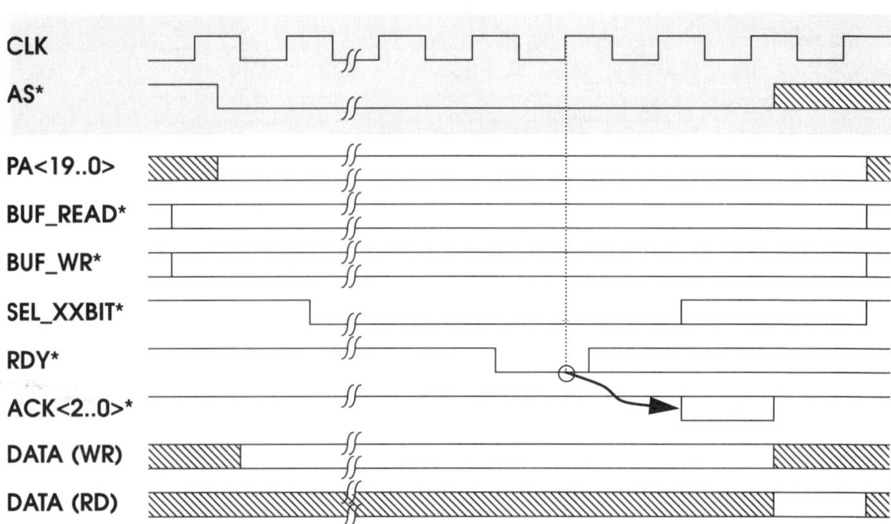

A variable number of wait-states may be added in the user-defined mode by asserting RDY* two clocks prior to desired ACK time.

FIGURE 9.13. *User-defined Prototype Interface timing (Reads or Writes).*

The timing for this mode of operation is shown in Figure 9.14. To generate an error acknowledgment instead of a normal data acknowledgment, the **ERR*** signal must be asserted for at least one full clock cycle. This signal will be synchronized by the SBus interface logic, and then used to initiate an error acknowledgment. Due to this synchronization, this signal must be activated at least two full clock cycles prior to the time a normal data acknowledge would otherwise have been generated. This requirement has the following consequences:

- Byte, half-word, or word operations with 0 wait-states cannot be terminated with an error acknowledgment via this mechanism.

- For transfers with four wait-states, **ERR*** should be asserted no later than the third clock-period after which **SEL_XXBIT*** becomes asserted.

- For transfers with 8 wait-states, **ERR*** should be asserted no later than the seventh clock-period after which **SEL_XXBIT*** becomes asserted.

- For user-timed transfers, **ERR*** should be asserted on or before the same clock that **RDY*** is asserted.

It is important to note that the error acknowledgment may not follow the **ERR*** signal by exactly two clocks. Due to limitations in the number of available terms in the PAL which generates the acknowledgments, the synchronized error signal is only sampled at key points in the wait-state counter. As a result, more than a two clock delay may occur, depending on the timing of **ERR*** and the number of wait-states in the current transfer.

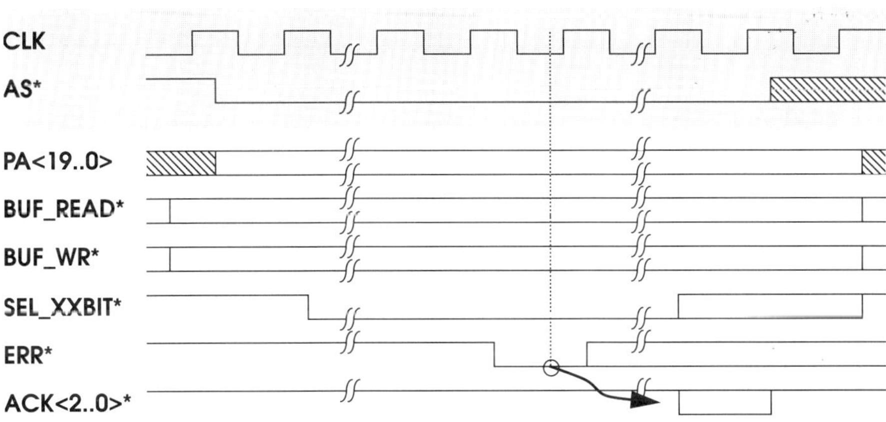

Error Acknowledgement will follow ERR* assertion by two or more clocks

FIGURE 9.14. *Terminating a cycle with an error.*

9.4.3 Burst Operations

Burst operations are virtually identical to the single-transfer operations described above, except that multiple words are transferred. Even this difference is of little importance in many cases, because from the prototype interface's perspective there is little difference between this and a series of separate transfers spaced closely together.

On burst read operations, the **SEL_32BIT*** signal goes and stays active through the transfer. The SERFboard will change the physical address after each segment is transferred. Though this does occur while the select line is asserted, it should be of little consequence in most cases. If the design contains registers whose contents are altered when read then care should be taken; as the address changes it may pass through "in-between" levels which could inadvertently alter such registers.

On burst write operations, the **SEL_32BIT*** line must be deasserted whenever the data or address changes. Once both address and data are stable and valid again, the **SEL_32BIT*** signal is reasserted. In this way, burst writes are exactly like a series of closely spaced transfers: any logic which the user adds need not know or care that a burst operation is in progress.

Burst operations are word-only operations. Any attempt to perform a burst operation in either the 8- or 16-bit address space of this card will result in unpredictable behavior.

9.4.4 Interrupts

There are provisions for a user specified interrupt, which is generated by asserting **INTA***. Upon detecting such a request the SERFboard's Prototype interface generates **INT_REQA***. This signal is a pseudo open-collector copy of **INTA***, and is buffered in such a way that it may be connected directly to the SBus interrupt signals **sb_irq(1:7)***. This is done via jumpers at location J3, which is diagrammed in Figure 9.15.

PIN	Signal	PIN	Signal
01	sb_irq6*	02	Int_reqa*
03	sb_irq5*	04	sb_irq7*
05	sb_irq3*	06	sb_irq4*
07	sb_irq1*	08	sb_irq2*

FIGURE 9.15. *Jumper Block J3 Pin-out.*

9.4.5 Power Considerations

The SBus specification limits single-slot SBus cards such as the SERFboard to 2 amps at 5V, and 30 mA at +12V and at -12V. This is a total for the entire card, which must take both the built-in interface and any circuitry added by the user.

As shipped, the SERFboard interface circuitry uses at most 1 amp at 5V and does not use +12V or -12V at all. This leaves the other 1 amp of the 5V current and all of the $\pm$ 12V. Remember that if you change any of the SERFboard's standard components (such as the PROM), these values may change.

Please evaluate your prototype design carefully, to guarantee that it stays within this power budget. Use the worst-case consumptions from the data books, too, or your best estimates of the worst-case consumptions for the particular environment.

If you find that even after careful consideration your design requires more current, all is not lost. If you have done a careful worst-case calculation and you are over by only a small amount, then it is probably okay to proceed with the prototype. Your design's "typical" current consumption may well be within reasonable bounds. Also, most hosts are designed using worst-case numbers, and statistically will usually be able to supply more current than the minimums they guarantee. If you choose to go ahead in this case, be on the lookout for problems that may be associated with a "droopy" power supply.

Other options include leaving one or more SBus slots empty while your prototype is installed. The current reserved for those slots would then be available to your board. Be careful, though, because too much current flowing through the SERFboard's power planes and associated connector pins can cause excessive voltage drops and ground offsets. This could cause intermittent operation and noise margin problems.

You might also consider providing an external power supply in extreme cases. This may be necessary if you must temporarily use parts that aren't CMOS, or if your design requires one or more of the plug-in boards that can be added to increase the usable prototype area.

Don't ever push the limits on a production design, however. It is always prudent to stay within the specification, and it is usually a good idea to build in at least a small margin. The real danger is that if your card pushes the limits "just a little," there could be real problems if it gets plugged into a system where either the host or another expansion card pushes on the same limit.

The power dissipation of your prototype may not be representative of the final design, but it is never too soon to start considering power issues before they become problems. In this way, possible solutions can be designed in when "an ounce of prevention is worth a pound of cure."

9.5 Schematics

The SERFboard schematics are shown in Figure 9.16. The components in the upper left corner, labeled U1, U2, and U3, are the address buffers for **PA(19:0)**. U1 and U2 are strictly buffers, but U3 is also a counter which increments the word address during burst operations. Notice that U3 buffers or modifies address bits **PA(2:5)**; **PA(0:1)** do not pass through the counter because they are not significant for word addresses.

Almost midway down along the schematic's left side is another buffer, labeled U8. This component buffers the Clock, **SIZ(2:0)**, **AS***, **SEL***, and **PA(23:24)**. Buffering the clock adds delay and skew to it, but negative side-effects are largely cancelled out because those signals whose timing relationship is most critical are buffered by the same part. Across any given single component, skews are minimized and delays are approximately identical. Therefore all signals buffered by the same part are delayed, but by about the same amount. The phase relationship between the signals is preserved.

The data transceivers are in the schematic's lower left corner, labeled U4, U5, U6, and U7. Internal to the SERFboard, all the data lines are pulled-up using 10 Kohm resistors in SIP packages. This is necessary to prevent these lines from floating during the time it takes for a peripheral to recognize its select signal and enable its output buffers. It is also necessary for those cases when the peripheral's width is less than the full 32-bit width the SERFboard provides.

The ID PROM (U10) is near the schematic's upper middle. Below it are the three control PALs, labeled U9, U11, and U12. U9 is also called the *ACKPAL*, because it manages access timing and drives the **Ack(2:0)*** signals. U11 is also called the *SELPAL*, because it performs some address decoding, qualifies it with **AS*** and **SEL***, and then generates the **SEL_XXBIT*** strobes. Finally, U12 is the *MISCPAL*, because it contains the miscellaneous small functions that would not fit into the other PALs. It contains some buffers and inverters, for example. It also contains the synchronizers for the **RDY*** and **ERR*** signals, as well as the pseudo open-collector driver for the **INT_REQA*** signal.

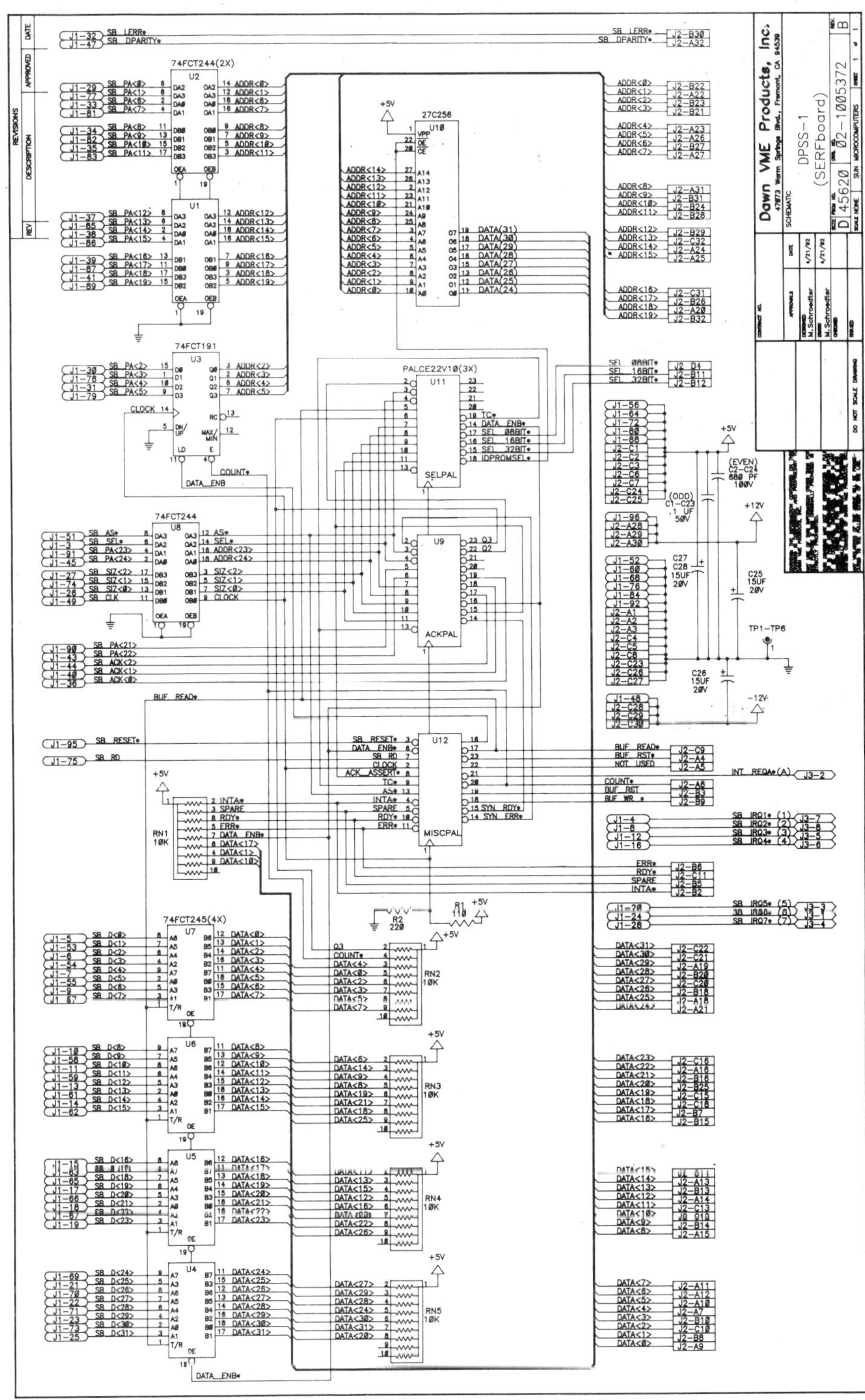

FIGURE 9.16. *SERFboard Schematics.*

The remainder of the schematic, all along the right side, is dominated by the connections to the J2 and J3 connectors. Also shown are the bypass capacitors and the (grounded) test points.

9.6 FCode and Software Drivers

The SERFboard is shipped with a 3.5" diskette which contains a variety of information. Among the files is one named **SERFcode.src**, which is the source code for the FCode contained in the board's ID PROM. Another file, named **driver.src**, is a sample driver. Please refer to the **read.me** file or the SERFboard's packing list for a description of the other files included on this diskette.

Generally, it should not be necessary to change the FCode much, if at all. The ID PROM's primary purpose is to identify the board when the system probes its SBus, and it is used to "attach" the appropriate driver during boot time. If changes are necessary, they most likely will be minor modifications to the **name**, **reg**, or **intr** attributes. For example, if the board is jumpered to interrupt at SBus level 4, the **intr** attribute will need to be modified accordingly. For information on how to accomplish this, please refer to *Writing FCode Programs for SBus Cards*, published by Sun Microsystems.

The sample software driver is another matter. It is highly unlikely that this driver will be of much use without substantial refinement. Drivers vary widely with the type of device, and the sample included is meant primarily as a basis upon which to build. For more information on driver related issues, please refer to *Writing SBus Device Drivers*, published by Sun Microsystems.

9.7 PAL Programs

It will occasionally be desirable to customize the programs of one or more PALs on the SERFboard. For example, you could modify the sizes of the three device selects so that you could have three separate 8-bit devices without subdividing the **SL_08BIT*** region. You might also wish to modify the PALs to further restrict the types of valid cycles; to exclude burst cycles, for instance. Another option might be to add support for rerun acknowledgments.

The information included in this section is offered to help understand how the existing PAL state machines function, so that any modifications needed can be done quickly and easily.

9.7.1 Source Files

The source file for the three PALs is included on the 3.5" diskette
that is shipped with the SERFboard. This file is named **PAL.SRC**,
and requires the "include" file named **PAL.INC**. This source file is
in a format understood by PLDesigner, a PLD development tool
which is a product of Minc Incorporated. The format is straightforward, however, and may be easily converted to a form understood
by ABEL, PALASM or many of the other popular PLD development
software packages.

For convenience, the PALs' JEDEC files can also be found on
this diskette. These are named **SELPAL.JED**, **ACKPAL.JED**, and
MISCPAL.JED, for the *SELPAL*, *ACKPAL*, and *MISCPAL*,
respectively.

9.7.2 Theory of Operations

There are several state machines and logic blocks that are divided
amongst the three PALs that form the heart of the SERFboard's
SBus interface.

The *SELPAL* contains two functions; the address decode logic
and the state machine which counts words in burst operations. The
address decoding is straightforward. The select signals are outputs
of combinatorial functions, and are gated so that they may be used
as strobes for either read or write operations (as discussed previously in the section titled 'Access Timing' starting on page 305. The
DATA_ENB* signal is also a combinatorial output of the address
decoding logic. As the name implies, this signal is used to enable
the SERFboard's data buffers. The state machine which counts
words during burst operations is basically a 4-bit counter. This
counter is pre-loaded with a value that is a function of the burst
size, and then counts down until the "terminal count" is reached.
There is one additional state bit which uniquely identifies this
state, and the SERFboard remains in this state until either **AS*** or
SEL* are de-asserted.

The *ACKPAL* contains the single state machine which controls
access timing and generates the data acknowledgments. Part of
this state machine is a 3-bit counter which inserts the desired number of wait-states in the cycle. This counter is pre-loaded with a
value that is a function of the address, or it may be bypassed if the
address indicates that the cycle time is to be externally controlled
(via the **RDY*** signal). The rest of this state machine is responsible
for driving the **SB_ACK(2:0)*** signals. Either an 8-, 16-, or 32-bit
acknowledgment may be generated (again a function of the

address), or an error acknowledgment will be generated (if the transfer requested is not supported or if the **ERR*** signal is asserted). In any case, the acknowledgment is asserted for exactly one clock, and then actively driven into the "off" state (as mandated by SBus' "active drive" requirement).

The *MISCPAL* provides a miscellaneous collection of functions. It contains buffers and inverters for a variety of the board's signals. It also contains synchronizers for both the **RDY*** and **ERR*** signals. It also provides a pseudo open-collector buffer for the **INTA*** signal, so that the result (**INT_REQA***) may drive one of the SBus' interrupt lines directly. Finally, there is a spare input and output (**SPARE** and **NOT_USED**, respectively) which may be user defined as described previously.

9.8 Application Examples

9.8.1 Interfacing an Intel 82C55A Programmable Peripheral Interface

Many possible applications require some form of parallel input and/or output. One example of such an application is block diagrammed in Figure 9.17. This design converts straight ASCII text into speech, using a dedicated co-processor which is programmed with algorithms based on work done by the Naval Research Laboratories.

The text-to-speech processor is not the focus of this application example, however. The emphasis here is on the 8-bit wide interface between this processor and the host CPU, and on the simple handshaking required (for dedicated control and status signals).

In this case the text-to-speech processor is interfaced to the SBus using the SERFboard and Intel's 82C55A Programmable Peripheral Interface chip. This chip's ancestry goes back a long way, and it is relatively slow, but it is extremely flexible and provides 24 bits of digital I/O in a variety of modes. It can be used to implement a wide variety of parallel interfaces. Please see Intel's data sheet on this part for a complete description, as well as examples of other possible applications. Some of these potential applications include parallel printer interfaces, motion and control applications, analog input or output, etc.

A detailed review of the 82C55A's data sheets and the SERFboard's timing characteristics shows that it may be simply interfaced to the SERFboard as long as at least four wait-states are used. The connections required are detailed in Figure 9.18. A 74FCT138

3-to-8 line decoder is used to sub-decode the 8-bit address space, and to gate the write and read strobes to the 82C55A. This last function is necessary because the 82C55A is a latch-based design that does not support chip-select strobed access timing.

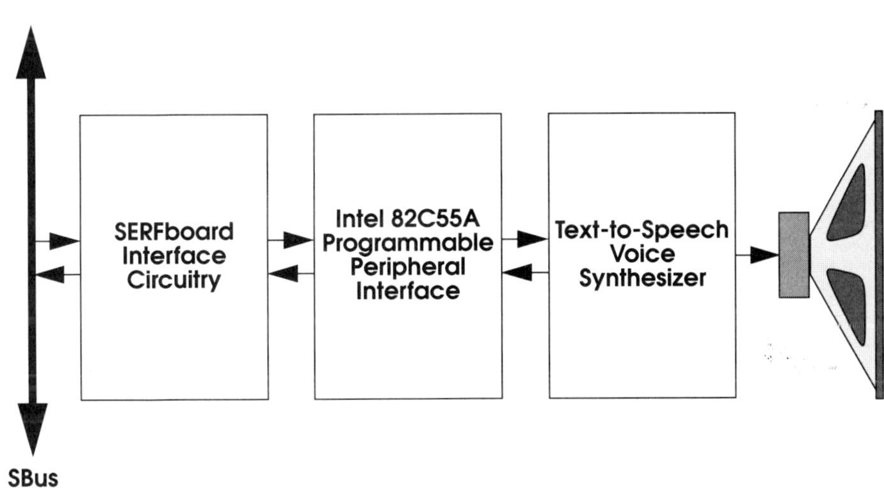

FIGURE 9.17. *Block Diagram of The VOICEboard.*

With this part interfaced to the SBus, we now have three distinct 8 bit ports that may be configured as inputs, outputs, or any combination. In the case of the VOICEboard, Port B is configured as an output port, and is used to set the text-to-speech processor's operating modes. Port A is configured as a tri-stateable output port, and is used to transfer ASCII characters to the processor.
Port C is used for miscellaneous inputs, interrupts, and data handshaking. The 82C55A is configured so that when data is written by the host CPU to Port A, an "output buffer full" flag is generated on one of PORT C's pins. This interrupts the text-to-speech processor which then reads the data by signaling an acknowledgment on yet another Port C pin. Interrupts may be generated when the data transfer has completed, also using Port C pins. Status information, too (such as the processor's "busy" flag or the output buffer full flag) can be read via Port C. The details of the interface between the 82C55A and the text-to-speech processor are not important here because they do not involve the SERFboard.

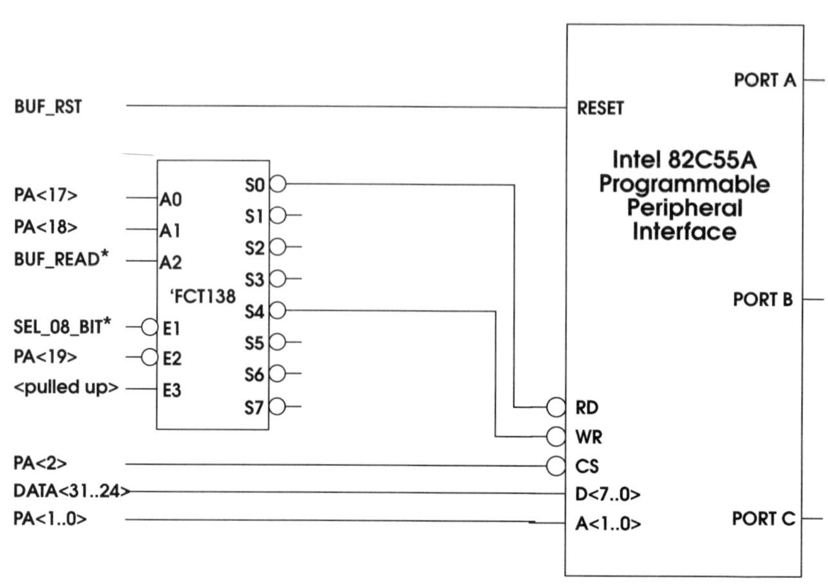

FIGURE 9.18. *Details of the 82C55A connection to the SERFboard.*

Other "microprocessor-compatible" parts can be interfaced to the SBus as easily as the 82C55A. This includes serial interfaces, analog-to-digital or digital-to-analog converters, motor controllers, co-processors, encryption units, etc. The SERFboard is an ideal solution for many applications where only a small number of large scale ICs need to be interfaced to an SBus-based host. This is true whether the application is a prototype of a commercial design, or if it is a case where only a few such 'products' are needed and a full scale development effort is not warranted.

9.9 Accessories and Information...

Dawn VME Products sells the SERFboard as Model DPSS-1, part number 06-1005372. Several accessories are available, including expansion panels which increase the amount of prototype area available. For more information about the SERFboard or its accessories, please contact:

Dawn VME Products
47073 Warm Springs Boulevard
Fremont, CA 94539
(415) 657-4444
(415) 657-3274 - Fax

The SBus Specification B.0, *Writing FCode Programs for SBus Cards*, *Writing SBus Device Drivers*, and other related documentation is available from a variety of distributors. Among them:

VME International Trade Association (VITA)
0229 N. Scottsdale Road, Suite B
Scottsdale, AZ 85353
(602) 951-8866

or:

VFEA
P.O. Box 192
5300 AD Zalthommel
The Netherlands
31-4180-14661

For information about PLDesigner, please contact:

Minc Incorporated
6755 Earl Drive
Colorado Springs, CO 80918
(719) 590-1155

Acknowledgments

Jim Lockwood and Jim Lyle co-designed the SERFboard.
Mike Saari and Deborah Bennet provided significant help with the sample
    FCode and sample drivers, respectively. Dave Caserza and Donn Dickerman
    of Dawn VME Products also provided substantial assistance.
ABEL is a trademark of Data I/O Corporation
PALASM is a trademark of Advanced Micro Devices, Inc.
PLDesigner and Minc are registered trademarks of Minc Incorporated.
SERFboard is a trademark of Dawn VME Products, Inc.

References

Sun Microsystems, Inc. *SBus Specification, Revision B.0*, January 1991

Glossary

Achieved Performance:	The best performance actually measured (or which one could expect to measure) in a real system. Achieved performance differs from the peak data rate because of differences in systems including clock rates, bus widths, the number of cycles required to access memory, the size of burst transfers supported, software and interrupt handling overhead, and transfers involving less than optimum sizes. Most of these factors are trade-offs made which depend on many factors, such as cost and time to market.
Asynchronous	Not synchronous (see *Synchronous* listing).
Atomic Transaction	A sequence of transfers during which one master has exclusive access. This prevents other transfers from disrupting a critical sequence, such as a read-modify-write operation.
Autoconfiguration	A process by which the system uses its own resources to determine, analyze, and configure the components present. This simplifies the end-user's task of customizing the machine for his or her own needs, because much of the intelligence required to perform this task is built-in.

Bandwidth	This is a measure of the rate at which data can be transferred over an interface. Bandwidth is usually a product of how much data can be transferred at any one time (interface width) and how often these transfers can occur (interface rate). In this text, bandwidth is measured in MB/s (Megabytes per second).
Boot Device	A device used for either input or output while the host's operating system is being loaded and configured. For example, the OS image may be fetched from a disk. Messages might be sent to a CRT display via a frame buffer, and "keyboard" commands may be read through a serial port. Therefore, these are all boot devices.
Burst Transfer	A single transfer in which multiple data words are read or written. This reduces the average protocol overhead associated with each word, improving overall efficiency.
Bus Bridge	An electrical and mechanical interface (or translator) that allows different kinds of buses to be connected together. One example would be an SBus to VMEbus translator, which allows VME cards to be accessed from an SBus-based host.
Bus-fight	A condition which arises in a bused environment when more than one output buffer attempts to drive the same signal at the same time. The situation is especially critical when the buffers try to drive the signal to different levels. Illegal voltage levels, excessive power dissipation, or signal ringing with large overshoots and undershoots may result. This could result in faulty operation or even component damage.

Byte	A collection of 8 data bits.
Byte Steering	The process by which data is multiplexed onto or off of the proper byte lanes. The byte lane(s) used by a transfer is (are) a function of the transfer's size, the slave's port width, and the address.
Component-Side	Opposite of solder-side. This term refers to the side of a printed-circuit board that historically held most of the board's components. The advent of surface mount technologies have blurred this definition because components can be mounted on both sides of the board. In this case the component-side is usually the primary side, and most often the top or front side of the board.
Clock Skew	The difference in the time at which a clock signal arrives at any two parts of a circuit.
CMOS	Abbreviation for *Complementary Metal-Oxide Semiconductor*.
Deadlock	Also called a deadly-embrace. This is a situation in which two or more processes compete for two or more resources. The problem arises when one process is granted access to a resource, and then can't give it up until another resource becomes available. At the same time, a different process owns the resource which is needed, and can't give that up until it gains access to the resource that the *first* process now owns. Neither process can proceed at this point.
DMA	*Direct Memory Access*. This is a feature that allows a device to transfer data to or from memory without requiring the help of the system's processor.

DVMA

Direct Virtual Memory Access. This is similar to DMA, except that a virtual address is used.

ESD

Electro-Static Discharge. A very high voltage pulse which results from the build-up of static electricity. It can damage integrated circuits if cautions are not taken to reduce or eliminate its effects.

EUROCARD

This is any of several standardized mechanical specifications for electronic cards established by the International Electrotechnical Commission (IEC). One prevalent example is the 233.35 x 160 mm "6U x 160" format often used by VME cards.

Expansion Box

In the SBus realm, an expansion box usually refers to a device which expands the number of SBus cards that can be connected to a host. The structure usually involves a single SBus card which plugs into the host. This card is connected via a cable to an external chassis which provides additional slots. This is a special form of bus bridge, and requires buffering logic in most cases to make it work effectively.

Falling-Edge

The slope or edge of a digital signal that occurs as it moves from an initial high value to an eventual low level.

FCode

A byte-tokenized representation of FORTH, which is a programming language that can be used to write self-test routines, simple "boot" device drivers, and so on.

Follow-on Transfers	Also called follow-on cycles. These are the SBus transfers generated to move the remainder of the data of an earlier, larger transfer that was bus-sized. For example, if a word transfer is given a byte acknowledgment, only the first byte will be transferred during that cycle. The remaining three bytes in the request must be moved using additional, follow-on, transfers.
Gate-Arrays	An ASIC in which the logic is composed of an array, or "sea," of simple gates. The designer specifies the interconnection of these gates to develop more complex functions.
Gator-Rays	The product of reptilian, phaser-toting aliens. This is what an engineer's spouse or non-technical friends think is being discussed when gate-arrays are mentioned.
Ground Bounce	A sudden change in ground potential which results from a sudden change in current flowing through the ground path's finite impedance. This can happen with chips, boards, or entire systems. A common example is a single integrated circuit which switches many of its outputs simultaneously. The internal ground reference of this chip may shift with respect to that of the other chips on the board. This degrades the system's noise margins, and is a primary reason to avoid ground bounce.
Half-word	In the context of the SBus, a group of two bytes, or 16 total data bits.
High-True	See *Positive-True*.

ID PROM	A read-only memory area that is required on each SBus card. It contains information which describes the card, its register and interrupt mappings, and the name of the device driver it uses. It may also contain FCode.
IEEE	*Institute* of *Electrical* and *Electronic* *Engineers*
"Interesting" Problems	Subtle, and usually very difficult to isolate. Often the initial symptoms seen have little to do with the actual problem, and can tend to lead the investigation astray. Very "interesting" problems are the major cause of hair loss and ulcers in both engineers and project managers, and are usually the result of short-cuts taken early in the design process.
Latch-up	A condition that CMOS circuitry is susceptible to, in which abnormal currents flow through some transistors in such a way that they cannot be switched normally. This condition might occur if an abnormally negative or excessively positive voltage is present at a device's input or output pins. Signal ringing is often a cause.
Latency	The time which elapses between the point at which an operation is initiated, and the point at which the operation completes. In the context of the SBus, this is the time between a master's first request for access to the bus, and the slaves final acknowledgment on the ensuing transfer.
Livelock	A situation that occurs when two or more processes attempt to simultaneously access a single resource in such

a way that they interfere. For example, a disk controller might be asked to perform a "seek" operation in preparation for a read. Before the read can be performed, however, a different seek request (in preparation for a different read) might occur. Then vice-versa. Despite the constant disk and request activity, no useful work is performed. (See also *Deadlock.*)

Long-word	In the context of the SBus, a group of eight bytes, or 64 total data bits.
Low-True	See *Negative-True.*
Master	A device that can initiate a data transfer (usually via DMA or DVMA) on a bus. Most masters also have slave capabilities.
Maximum Bandwidth	The maximum rate at which data can be transferred; calculated as the product of the transfer rate and the transfer width.
MOSFET	Abbreviation for *Metal-Oxide-Silicon Field Effect Transistor.*
Motherboard	The circuit board on which the main circuitry of a system is mounted, and to which secondary, or add-on, boards can be connected.
MTBF	*Mean Time Between Failures*
"Natural" Location	Used in reference to a byte's location on the **Data** lines, when it is aligned with the byte's address. For example, the natural location of byte 0 is **D(31:24)** (byte lane 0), that of byte 1 is **D(23:16)** (byte lane 1), and so on.

Negative-True	A reference to a signal's polarity. When a negative-true signal is asserted its measured voltage will be more negative than the receiver's threshold.
P1275	The designation of the effort within the IEEE to produce and Open Boot PROM and FCode Standard compatible with that organization's policies and requirements. When available, that standard will be designated 1275.
P1496	The designation of the effort within the IEEE to produce an SBus Standard compatible with that organization's policies and requirements. When available, that standard will be designated 1496.
Peak Data Rate	The best performance achieved and sustained within the specified limits of the interface. Peak data rate includes protocol overhead and thereby differs from the maximum bandwidth. Note: Digital Equipment Corporation's literature refers to this as "Protocol Performance."
Pm	Symbol for Promethium.
Positive-True	A reference to a signal's polarity. When a positive-true signal is asserted its measured voltage will be more positive than the signal's threshold.
Rising Edge	The slope or edge of a digital signal that occurs as it moves from an initial low value to an eventual high level. All synchronous SBus times are measured from the rising edge of the clock.
Sample	To record the state of a signal at a narrow, pre-determined time. Usually at an edge of a clock signal.

Slave

A device that can respond to a data transfer on a bus, but cannot initiate one.

Solder-Side

Opposite of component-side. This term refers to the side of a printed-circuit board that historically did not usually contain components. Instead, this is the side through which the component leads projected, and to which solder was applied. The advent of surface mount technologies has blurred this definition because components can be mounted on both sides of the board. In this case the solder-side is usually the secondary side, and most often the bottom or back side of the board.

SMT

Surface Mount Technology.

State Machine

Also called a sequential circuit. A digital logic circuit which contains a core whose internal values step from one pre-defined *state* to another, usually at fixed intervals defined by a clock signal. Both the circuit's outputs and its next state may be a function of the circuit's inputs and current state.

Synchronous

Operation in step (or in phase) with some reference. Within digital logic circuitry, this specifically refers to a flip-flop or state-machine which accepts inputs only at fixed time intervals defined by a clock signal.

Time out

A situation in which a device does not respond within a specified time limit. Usually this is the result of an error or failure, and the transaction must be terminated. On the SBus the controller is responsible for doing this, by signalling an error acknowledgment.

VME	VERSAmodule Eurocard. Also known as the VMEbus, this is a standardized backplane interface that evolved from Motorola's VERSAbus, and has gained wide market acceptance.
Wait-State	An extra state or clock added because a result is not ready yet. The effect is to delay or "stretch" an operation when necessary. Usually this is done to synchronize fast devices with slower ones.
Word	In the context of the SBus, a group of four bytes, or 32 total data bits.

Bibliography

Barnes, J. R. *Electronic System Design Interference and Noise Control Techniques*. Prentice-Hall, 1987.

Di Giacomo, J. ed. *Digital Bus Handbook*. McGraw Hill, 1990.

Digital Equipment Corporation. "DECstation and DECsystem 5000 Model 200 Technical Overview."

Digital Equipment Corporation. "TURBOchannel Developer's Kit - Version."

Digital Equipment Corporation. "TURBOchannel Overview." April 1990.

Glass, B. "Power Management." *BYTE*, September, 1991.

Glass, L. B. "Inside EISA." *BYTE*, November 1989.

Honan, P. "EISA under the Glass." *Personal Computing*, August 1989.

IBM Corporation. *IBM Micro Channel Supplement for the PS/2 Hardware Interface Technical Reference*. November 1989.

IEEE. IEEE P896.1 (Futurebus+ Logical Layer Specifications). Draft 8.1a.

IEEE. IEEE P896.2 (Futurebus+ Physical Layer and Profiles). Draft 4.0.

IEEE. IEEE P1101.2 (Mechanical Core Specifications for Microcomputers Using 2 mm Pitch Connectors). Revision 0.

IEEE. IEEE P1275 (Open Boot Draft Standard). Revision 02.

IEEE. IEEE P1496 (SBus Draft Standard). Revision 02.

Lewis, P. H. "IBM turns back the clock." *New York Times*. (appeared in the *San Jose Mercury News. Computing Section*. June 16, 1991).

Lyle, J., M. Sodos, and S. Carrie, "The SBus: Designed and Optimized for the User, the Developer, and the Manager." *BUSCON-East Conference Proceedings*, October 1990

Ott, H. W. *Noise Reduction Techniques In Electronic Systems*. John Wiley and Sons, 1976.

Perry, D. "Multiple Technologies Shrink Future PCB Modules." *Printed Circuit Design*, February, 1991.

SDS Data Systems. *Grounding and Noise Reduction Practices*.

SPARC International, *SPARC-Line (Newsletter)*, September 1991.

Sun Microsystems, Inc. *SBulletin*, June 1990

Sun Microsystems, Inc., *SBus Application Notes*, No.1, Revision A, July 1990.

Sun Microsystems, *Writing FCode Programs for SBus Cards*, 1990.

Sun Microsystems, Inc. *SBus Specification, Revision B.0*, January 1991

Sun Microsystems, Inc., *SBulletin*, March 1991

Turner, R. P. and Gibilisco, S. *The Illustrated Dictionary of Electronics*. TAB Books, 1988.

Index